Springer Theses

Recognizing Outstanding Ph.D. Research

For further volumes:
http://www.springer.com/series/8790

Aims and Scope

The series "Springer Theses" brings together a selection of the very best Ph.D. theses from around the world and across the physical sciences. Nominated and endorsed by two recognized specialists, each published volume has been selected for its scientific excellence and the high impact of its contents for the pertinent field of research. For greater accessibility to non-specialists, the published versions include an extended introduction, as well as a foreword by the student's supervisor explaining the special relevance of the work for the field. As a whole, the series will provide a valuable resource both for newcomers to the research fields described, and for other scientists seeking detailed background information on special questions. Finally, it provides an accredited documentation of the valuable contributions made by today's younger generation of scientists.

Theses are accepted into the series by invited nomination only and must fulfill all of the following criteria

- They must be written in good English.
- The topic should fall within the confines of Chemistry, Physics, Earth Sciences, Engineering and related interdisciplinary fields such as Materials, Nanoscience, Chemical Engineering, Complex Systems and Biophysics.
- The work reported in the thesis must represent a significant scientific advance.
- If the thesis includes previously published material, permission to reproduce this must be gained from the respective copyright holder.
- They must have been examined and passed during the 12 months prior to nomination.
- Each thesis should include a foreword by the supervisor outlining the significance of its content.
- The theses should have a clearly defined structure including an introduction accessible to scientists not expert in that particular field.

Aleksandar Borisavljević

Limits, Modeling and Design of High-Speed Permanent Magnet Machines

Doctoral Thesis accepted by
Delft University of Technology, The Netherlands

 Springer

Author
Dr. Aleksandar Borisavljević
Department of Electrical Engineering
Eindhoven University of Technology
Eindhoven
The Netherlands

Supervisors
Prof. Dr-Eng. J. A. Ferreira
Faculty of Electrical Engineering,
   Mathematics and Computer Science
Delft University of Technology
Delft
The Netherlands

Dr.ir Henk Polinder
Faculty of Electrical Engineering,
   Mathematics and Computer Science
Delft University of Technology
Delft
The Netherlands

ISSN 2190-5053   ISSN 2190-5061   (electronic)
ISBN 978-3-642-33456-6   ISBN 978-3-642-33457-3   (eBook)
DOI 10.1007/978-3-642-33457-3
Springer Heidelberg New York Dordrecht London

Library of Congress Control Number: 2012947858

Printed on acid-free paper

Springer is part of Springer Science+Business Media (www.springer.com)

Parts of this thesis have been published in the following academic articles:

Aleksandar Borisavljevic, Henk Polinder, Jan A. Ferreira: "On the Speed Limits of Permanent Magnet Machines", IEEE Transactions on Industrial Electronics, January 2010, Vol. 57, No. 1 pp. 220–227.

A. Borisavljevic, H. Polinder, J.A. Ferreira: "Conductor Optimization for Slotless PM Machines", Proceedings of the XV International Symposium on Electromagnetic Fields in Mechatronics, ISEF 2011.

A. Borisavljevic, H. Polinder, J.A. Ferreira: "Calculation of Unbalanced Magnetic Force in Slotless PM Machines", Proceedings of Electrimacs, 2011.

A. Borisavljevic, H. Polinder, J.A. Ferreira: "Enclosure Design for a High-Speed Permanent Magnet Rotor", Proceeding of the Power Electronics, Machines and Drives Conference (PEMD), April 2010, Vol. 2, pp. 817–822.

A. Borisavljevic, H. Polinder, J.A. Ferreira: "Realization of the I/f Control Method for a High-Speed Permanent Magnet Motor", Proceedings of the International Conference on Electrical Machines (ICEM), September 2010, Vol. 4, pp. 2895–2900.

A. Borisavljevic, M. H. Kimman, P. Tsigkourakos, H. Polinder, H. H. Langen, R. Munnig Schmidt, J.A. Ferreira: "Motor Drive for a Novel High-Speed Micro-Milling Spindle", Proceedings of the International Conference on Advanced Intelligent Mechatronics (AIM) 2009, pp. 1492–1497.

to my lovely parents

Supervisors' Foreword

It is with great pleasure that we introduce Dr. Aleksandar Borisavljević's thesis, an outstanding work in the field of high-speed electric machines, completed at Delft University of Technology. His work focuses on the analysis and design of high-speed permanent magnet machines, based on the test case of a high-speed spindle drive for micro-machining. High-speed machines are gaining prominence and are increasingly used in applications such as micro-gas turbines, turbo-compressors, turbo-molecular vacuum pumps, generator-motor units for flywheels, and electrically driven turbo-chargers.

What sets this work apart is Dr. Borisavljević's broad approach to the research, having chosen to analyze different limits of high-speed PM machines, as well as their modeling and design. The modeling in the thesis represents important electromagnetic, structural, and rotordynamical aspects of these machines using both analytical and finite element methods.

Although broad, his approach does not sacrifice thoroughness or depth. After derivation, these models are used in the specific design of a small high-speed motor supported by static air bearings. In addition, the thesis presents a controller used for stable control of the high-speed motor fed by a voltage-source inverter. The design is finally tested and the experimental results are used to validate the models derived in the earlier chapters.

The thesis concludes with a critical assessment of the models, design, and suitability of the chosen machine and bearing type for high-speed applications.

Delft, The Netherlands, August 2012   Henk Polinder
                                                            Prof. Dr-Eng. J. A. Ferreira

Acknowledgments

In September 2006, I moved to The Netherlands and began my Ph.D. project in the Electrical Power Processing (EPP) group of TU Delft. Both working on a Ph.D. project and living in the Netherlands have helped me gain valuable personal and professional experience. Today, I look back with content on the past five years that enriched me as a person more than any other period in my life. But foremost, I feel I owe gratitude to many wonderful people that I was fortunate to become acquainted with, both personally and professionally.

First, I would like to express my gratitude to my co-promoter and daily supervisor, Henk Polinder, whose supervision, support and friendship were crucial for me to persevere and overcome the challenges of doctoral work. I feel privileged to have had such a wonderful person as a supervisor and true friend.

To my promoter, Prof. Braham Ferreira, I am sincerely grateful for helping me manage my Ph.D. work. Without our monthly meetings and his brilliant feedback, I would have never been able to find my way in engineering research.

I would like to thank Jelena and Mark Gerber for their support and, especially, for taking care of me like a family when I first moved to Delft. Their encouragement and friendship were such an important influence, allowing me to quickly adapt and enjoy life in my new environment.

Everyone who has ever done something in the EPP lab knows what Rob Schoevaars means to the group and Ph.D. students. His experience, patience, and good spirit helped me to overcome all those stubborn practical problems one must always face when doing experiments.

I am most grateful to the secretary of the department, Suzy Sirks-Bong, who has always been ready to go out of her way to help me and other Ph.D. students with *any* kind of problem.

I would like to express my gratitude to all the people of the Microfactory project team for their collaboration and, in particular, to Hans Langen and Maarten Kimman whose work and ideas formed the basis of my Ph.D. research. I would also like to thank Prof. Rob Munnig Schmidt—discussions with him were always inspiring.

This thesis would certainly not be the same without Petros Tsigkourakos, who designed the air-bearings setup and inverter which I used to test my designs. It was a very demanding assignment, but Petros worked relentlessly until we had a working setup. Thanks to his diligence, we managed to overcome many problems and both learned a lot.

A university is a very vibrant intellectual environment, and the feedback and advice of many people have had an influence on my work. I must distinguish the great help I received from my colleague and office-mate, Deok-Je Bang, whose vast experience and brilliant ideas helped me to work out the mechanical aspects of my designs. His ideas are interwoven into many solutions presented in this thesis.

I would also like to thank Zhihui Yuan for his help in DSP programming, Navin Balini for advice on control and Domenico Lahaye for his assistance with finite element modeling. Dr. Frank Taubner from Rosseta GmbH gave very valuable input to the design of the rotor retaining sleeve.

It was a great pleasure working in the EPP group, I am happy that we established such a friendly atmosphere among the Ph.D. students, enhanced by both mutual support and fun. I was particularly lucky to share an office with such great guys: Deok-Je Bang, Frank van der Pijl, and Balazs Czech. Our office was always known for its homelike atmosphere.

My wife, Veronica Pišorn, improved my academic writing immensely and spent long hours in editing my papers. I learned from her how the good and professional presentation of ideas can improve scientific work.

In Delft, I met a lot of amazing people and gained very close friends. It is privilege to be surrounded by so many interesting and smart people. I am particularly proud of my friends from ex-Yugoslavia who reflect the best of the culture and spirit of our turbulent region. I am thankful for their love and support throughout all these years.

I am very happy that I volunteered in Filmhuis Lumen, one of the nicest places in Delft. Hanging out with cozy and interesting people from Lumen helped me learn the language and get to know with Dutch people, along with doing something that I genuinely like.

Finally, I would like to mention those people who have personally meant the most to me. To them I owe my deepest gratitude.

To Jasmina and Darko for sharing their lives with me, for their love and support, for giving me ease when I needed it the most.

To Nada, Zoran and Nemanja, my wonderful family, for their love and dedication, for being so sincere, for showing how truth and honesty is invincible.

To Veronica, love of my life, for her support and optimism, for unfolding so much beauty, taste and joy in everyday.

And to my dear friends and family in Serbia, who I carry in my heart wherever I go.

Eindhoven, July 2011   Aleksandar Borisavljević

Contents

Symbols and Abbreviations

Latin Letters

$A, \vec{A}$	Magnetic vector potential	(Wb m^{-1})
A	Surface area	(m^2)
A_c	Current loading	(A m^{-1})
$B, \vec{B}$	Magnetic flux density	(T)
B_{ij}	Compliance matrix (element)	(N m^{-1})
b	Width	(m)
b	Compliance	(N m^{-1})
bei	Kelvin function (imaginary part)	
ber	Kelvin function (real part)	
C	Capacitor, Capacitance	(F)
C	Iron-loss coefficient	(*various*)
C_f	Air friction coefficient	()
C_i	Boundary coefficient	(T)
c	Coefficient of viscous damping	(N s m^{-1})
d	Diameter	(m)
D_i	Boundary coefficient	(T m)
E, e	No-load voltage	(V)
E	Young's modulus	(Pa)
$F, \vec{F}$	Force	(N)
$F()$	Function	
$\mathcal{F}, \mathcal{G}, \mathcal{H}$	Positive two-dimensional stress functions	(Ps)
F_d	Force density	(N m^{-2})
f	Frequency	(Hz)
G	Shear modulus	(Pa)
G_{ij}	Gyroscopic matrix (element)	(kg m^2)
$G()$	Function	
g	Effective air gap	(m)
$H, \vec{H}$	Magnetic field intensity	(A m)

h	Height	(m)
I, i	Electrical current	(A)
I	Surface moment of inertia	(m^4)
J, j	Electrical current density	(A m^{-2})
J	Moment of inertia	(kg m^2)
K_{ij}	Stiffness matrix (element)	(N m^{-1})
k	Portion of frequency-dependent losses in total copper loss	()
k	Roughness coefficient	()
k	Stiffness	(N m^{-1})
L	Inductance	(H)
l	Length	(m)
M_{ij}	Inertia matrix (element)	(kg)
m	Mass	(kg)
N	Number of winding turns	(turns)
n	Angular conductor density	(turns rad^{-1})
n	Number of conductor strands	()
P, p	Real power	(W)
p	Power density	(W m^{-3})
p	Static pressure	(Pa)
q	Displacement function	(m)
R	Resistance	(Ω)
Re	Reynolds number	()
S	Surface area	(m^2)
r	Radius	(m)
T	Torque	(N m)
T	Temperature	(°C, K)
Ta	Taylor number	()
t	Time moment	(s)
U, u	Voltage	(V)
u	Displacement	(m)
V	Volume	(m^3)
v	Velocity	(m s^{-1})
W	Weight	(kg)
w	Width	(m)
x	x-coordinate	
y	y-coordinate	
z	z-coordinate	

Greek Letters

α	Ratio between active and total rotor length	()
α	Coefficient of linear thermal expansion	(K^{-1})
Γ	Gyroscopic moment, non-dimensional	()
δ	Skin depth	(m)

δ	Clearance	(m)
ε	Rotor eccentricity	()
ε	Strain	()
ε	Static unbalance	(m)
θ	Angle	(rad)
λ	Rotor slenderness	()
μ	Magnetic permeability	(H m^{-1})
μ	Dynamic viscosity	(kg m^{-1} s^{-1})
ν	Poisson's ratio	()
ρ	Mass density	(kg m^{-3})
ρ	Electrical resistivity	(Ω m)
σ	Electrical conductivity	(S m^{-1})
σ	Stress	(Pa)
σ_M	Maxwell's stress	(Pa)
τ	Temperature increment	(K)
Φ	Magnetic flux	(Wb)
ϕ	Ratio between conductor diameter and the skin-depth	()
φ	Angular position	(rad)
χ	Shear parameter	()
χ	Couple unbalance	(rad)
ψ	Flux linkage	(Wb)
Ω	Angular velocity	(rad s^{-1})
ω	Angular frequency	(rad s^{-1})

Latin Subscripts

a, b, c	Phase
AC, ac	Alternated current—ac
BH	BH curve
c	Current
$cr, crit$	Critical value
Cu	Copper
DC, dc	Direct current—dc
$diss$	Power dissipation
e	Electrical
e	Sleeve/enclosure
$e, eddy$	Eddy currents
eff	Effective value
ei	Sleeve/enclosure—inner
eo	Sleeve/enclosure—outer
f	(Air) Friction
Fe	Iron
g	Air gap
h	Hysteresis

lin	Linearized
m	Magnet
mag	Magnetic
max	Maximum or limiting value
mech	Mechanical
n	Non-rotating
nom	Nominal (rated) value
opt	Optimal
PM	Permanent magnet
p	Polar
prox	Proximity effect
r	Rotor, rotating
r	Radial coordinate/direction
rec	Recoil
ref	Reference
rem	Remanent
res	Resonant
s	Stator
s	Slot
skin	Skin effect
so	Stator outer
st	Strand
t	Tooth
t	Tangential coordinate/direction
t	Transversal
U	Ultimate or yield
vM	Von Mises
w	Winding
wc	Winding center
x, y, z	Dimensions in Cartesian coordinate system
z	Axial coordinate/direction
0	Air/vacuum

Greek Subscripts

θ	Tangential coordinate/direction
φ	Tangential coordinate/direction

Abbreviations

AMB	Active magnetic bearings
CHP	Combined heat and power
CTE	Coefficient of thermal expansion
DN	DN number: product of bearing diameter and rotational speed

EDM	Electro-discharge machining
EM	Electromagnetic
FE	Finite element
FEM	Finite element model
MEMS	Microelectromechanical systems
PM	Permanent magnet
RMS	Root mean square
1D	One-dimensional
2D	Two-dimensional
3D	Three-dimensional

Vectors

$\vec{i}_r$, $\vec{i}_\varphi$, $\vec{i}_z$ Unity vectors of r, φ and z axes in Cylindrical coordinate system

Chapter 1
Introduction

1.1 Machinery for Micromilling

Miniature, micro/mesoscale components with 3D features are in demand for various industries such as electro-optics, biomedical, electronics, aerospace, etc. [1]. Applications for which micro-products are needed include optical molds and assembly, medical diagnostic devices, medical implants, electronic devices and chemical micro-reactors [2]. The microsystem technology market has grown steadily [3, 4] and with it the demand for high accuracy, complex shapes and a broader assortment of materials.

Different fabrication processes have been used for the production of miniature components in large batches: chemical etching, photo-lithography, laser, ultrasonic and ion-beam cutting and electro discharge machining. A majority of these methods is limited to silicon-based materials and/or planar geometries [1, 5]. For obtaining essentially 3D components, with dimensions ranging from a few hundred microns to a few millimeters and with micron-sized features, micro-mechanical machining has become the most viable solution. This kind of machining represents, in effect, the scaling down of traditional cutting technologies such as milling, drilling and grinding (Figs. 1.1 and 1.2). By accommodating conventional macro processes on a micro scale, we are enabled to produce even small batches of micro components at lower costs and in a wide range of materials: metallic alloys, ceramics, glass etc [4]. In this way, micromachining bridges the gap between silicon-based MEMS processes and conventional precision machining [6].

Very strict requirements—in terms of high structural stiffness and damping, low thermal distortion, environmental isolation—for the mechanical structure of micro-scale machinery [4, 6] have caused commercial machines be extremely large and heavy [7]. Their high cost and limited flexibility bring about high initial cost and limit them to a narrow range of applications that require a large number of high-precision components with relatively simple geometries. As stated in [2]: "Despite all of the progress in mechanical machining (...), the fact that it often takes a two-ton machine tool to fabricate microparts where cutting forces are in the milli- to micro-

A. Borisavljević, *Limits, Modeling and Design of High-Speed Permanent Magnet Machines*,   1
Springer Theses, DOI: 10.1007/978-3-642-33457-3_1,
© Springer-Verlag Berlin Heidelberg 2013

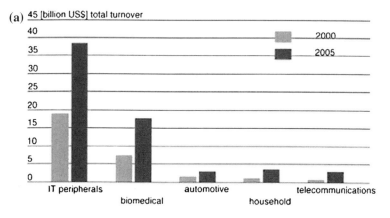

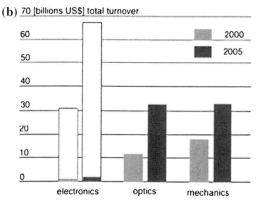

Fig. 1.1 Application market of microproducts **a** by industry; **b** by discipline (taken from [4])

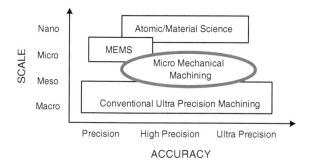

Fig. 1.2 Dimensional size for the micro-mechanical machining (according to [1])

Newton range is a clear indication that a complete machine tool redesign is required for the fabrication of micromachines."

"Small Equipment for Small Parts" [2] is a good paradigm for the global trend towards the miniaturization of manufacturing equipment and systems for micro-scale products. Many academic institutions, mainly in Europe and Asia, have been directing efforts toward the development and commercialization of smaller, even desktop-sized micro-cutting machines. These efforts have resulted in a number of compact, three to five-axis, mainly micromilling machines [7–17].

Important advantages of small micro-cutting machines can be stated as follows [7, 8]:

- small footprint and weight;
- ease of localized environmental and safety control;
- energy efficiency;
- portability and reduced transportation costs;
- reduced initial investment costs and costs from energy consumption and maintenance;
- a full spectrum of micro-machining applications;
- higher speed and acceleration due to smaller spindle size and inertia;
- positing effect on monitoring of the cutting process [18];
- shortened ramp-up process in production;
- allowing machine modular design and reconfiguration;

Two major trends towards machine miniaturization could be distinguished. In Japan and, recently, Korea the "Microfactory" concept has been dominant according to which one or more small machines are placed on a desktop to produce microparts in a fully automated process [8, 15]. However, microfactories did not offer a major contribution in the fabrication of microparts or final accuracy and most related projects have remained in the research phase [7, 19].

Most European projects, on the other hand, have been focusing on machine and process [2]. This approach practically represents a trade-off between large conventional machines and microfactories: the machinery is downsized and flexibility in producing complex 3D parts is reached, while stiffness and accuracy of the precision machines is maintained. A typical representative of this concept is Ultramill—a bench-top 5-axis micromilling machine developed in Brunel University, UK [7].

The spindle is a key part of micro-cutting machinery since it has the most significant effect on the quality of the machined components [4]. The spindle needs to have high motion accuracy and be capable of operating at very high-speeds. The dynamic of the rotor-bearings system is very important: when machining with micro-scale tools, small spindle deviations can result in large run-out [19].

In micromilling the speed requirement becomes particularly severe. As sizes of the features to be machined decrease, the diameters of the required tools decrease accordingly. In order to achieve cutting speeds which are standard for macroscopic milling tools, extremely high rotational speeds of a spindle are needed [20]. For instance, for milling materials such as stainless steel or brass with a $50 \mu m$ tool, required spindle speeds are beyond 380.000 and 950.000 rpm respectively [21]. At the moment of writing, speeds of machining spindles do not exceed 200.000 rpm and advanced spindle technologies are expected to support higher speeds in future.

Most existing small and medium size high-speed spindles use aerostatic bearings [4]. Because of their very small air-gaps (in the order of tens of μm) air bearings require flawless geometry with very low tolerances. Extremely high motion accuracy can therefore be achieved with high precision and repeatability. Compared to their dynamic counterpart aerostatic bearings consume less energy and can efficiently operate in a much wider speed range [7].

Active magnetic bearings (AMB) represent another promising technology for micro-machining spindles. AMB offer a possibility of creating practically arbitrary damping or stiffness that can be tailored to the operating mode (speed) [22]. In spite of these and a number of other advantages—such as high load capacity, no inherent instabilities, modularity—AMB spindles are still in the research stage with regard to micro-machining applications.

Motorized spindles are suitable for high-speed rotation. Such a design eliminates the need for power-transmission devices [23] and offers higher torque and efficiency than in the case of air-turbine-driven spindles [1, 7]. Incorporating a motor into the spindle reduces its size and rotor vibrations [4, 23], at the same time complicating thermal and mechanical design and modeling [23]. When defining important elements of the spindle—bearings, rotor, motor—a holistic approach to design may be preferred [7].

Regarding types of built-in spindle motors, mostly permanent magnet (PM) machines have been used [7]. PM machines are well-suited for very-high-speed applications due to high efficiency and high power density, particularly in low volumes [24, 25]. Conventionally, electrical machines are optimized according to output torque/power requirement and permissible heat dissipation which are result of their electromagnetically-induced losses. In miniature machining spindles, however, motor design and optimization becomes closely connected with the design of other key elements of the spindle.

1.2 High-Speed Spindle Drive

Using the Japanese "micro-factory" as an inspiration and paradigm, the Dutch Microfactory project centers its research on process and assembly technology for the fabrication of 3D products on a micro scale. The focus is on downsized equipment for micro-structuring, -assembly, -sensing and -testing. The Microfactory project [26] virtually combines the ongoing research in micro-systems technologies with the "micro-factory" concept: the researched machinery is desktop-sized and different process techniques are even intended to be combined on the same device (e.g. high-speed cutting and EDM).

Within the Microfactory project, the study in microstructuring is mostly concerned with micromilling as the most promising technology for production of 3D micro-parts. In order to make milling technology competitive for the micro-scale products, micromilling requires spindles with speeds beyond 300.000 rpm and a positioning accuracy in the order of 1 μm. New specialized spindles with small rotors and frictionless—magnetic or air—bearings are necessary to reach such rotational speeds and accuracy.

Spindles with built-in motors facilitate precise control of rotor acceleration and deceleration. Furthermore, the inclusion of a motor allows for a compact machine, as explained in the previous section. However, spindle drives that support extremely high rotational speed and are compatible with frictionless, low-stiffness bearings are not readily available on the market. Electrical drives specially designed for high-speed machining spindles motivated work on this thesis.

Recently, very high-speed machines have been developed for applications such as microturbine generators, turbocompressors and aircraft secondary power systems. In these examples, the demand for compact and efficient electromechanical systems has created the necessity of high-speed machine use [27]. The elimination of gears through the use of direct-driven electrical machines has resulted in improved system reliability.

Permanent magnet machines have become prevalent in low volume applications [24]. This could be ascribed to their magnetic excitation—the air-gap flux density of PM machines is determined predominantly by the quality of utilized permanent magnets and does not depend on the size of the machine. Several extremely high-speed PM machines have been reported, both on the market and in academia, with speeds up to 1 million rpm [28–30]. These relatively simple machines are proposed for direct-drive generators; the inclusion of a motor in a machining spindle, however, brings about new challenges in the area of electrical machines.

The advance in speed of spindle motors is linked with overcoming or avoiding a number of machine limitations. Various physical parameters (stress, temperature, resonant frequencies) can limit the speed of an electrical machine. Aside from speed, these variables are also affected by power, size and electrical and magnetic loading. They need to be carefully designed so that the spindle temperature, structural integrity and rotordynamic stability are not jeopardized at the operating speed [25].

Additional challenges lie in the control of high-speed machines. Due to limitations of the processing powers of microcontrollers, very high-speed machines are usually controlled in open-loop [29]. However, the open-loop speed response of PM machines becomes unstable after exceeding a certain speed [31], therefore, special control algorithms are required that would ensure stability without a great computational burden.

1.3 Problem Description

The project commenced in a research group within the Microfactory project. The group has worked towards new desktop machinery for micromilling, focusing on the spindle as the machine's key element. Active magnetic bearings (AMB) were recognized as a particularly attractive technology for micromilling spindles. Beside frictionless operation, AMB facilitate supercritical operation and exploiting self-centering of the rotor thus, with a reduction in the rotor size, desired high rotational speeds could be achieved. Yet, a new electrical drive for the spindle was needed.

The project goal was to design and build a high-speed electrical drive which would be compatible with a soft-mounted spindle for speeds beyond 300.000 rpm. Hence, it was primarily expected that the spindle motor has low stiffness (unbalanced pull), generates minimum frequency-dependent losses (heat) and that the motor design does not compromise the strength and robustness of the rotor.

In the initial phase of the project, a permanent magnet machine was chosen as the best candidate for low-volume high-speed applications. To minimize losses in the rotor, a synchronous type of PM machine with sinusoidal currents in the stator was chosen. That implied the necessity of choosing or designing a power inverter that would supply the motor with sinusoidal currents and, preferably, facilitate development of an adequate motor controller.

However, it was apparent that the application requires more than a conventional electromagnetic design of an electrical machine and its drive. Structural and rotor-dynamical aspects imposed equally great challenges to the design. The reduction in size of the rotor would also demand a closer integration of the motor and bearings and, in turn, their higher mutual dependence. New solutions were sought among unconventional spindle concepts [32].

At the same time, the project gave incentive for scientific research in the field of high-speed machines and, particularly, for the modeling and design of high-speed PM motors. The intention was to identify phenomena, both mechanical and electromagnetic, that take precedence in high-speed machines. By doing that, it was possible to single out a suitable motor configuration (in terms of number of poles, winding configuration, materials, etc.). Moreover, an analytical representation of those phenomena would enable the design and optimization of very high-speed machines including a spindle motor.

The demand for such extremely high speeds spurred an inquiry into the limitations of the machines: what does limit the speed of current machinery? Could those limits be overcome or avoided? This study implicitly formed an approach for designing machines for high-speed applications: (i) recognize the speed limits of the machines, (ii) correlate the limits with machine parameters and (iii) overcome those limits with new solutions which are adequate for a particular application.

Eventually, new concepts and models needed validation in a practical demonstrator/setup. It was envisaged that the setup would be used for testing the spindle drive and its suitability for the application as well as verifying or refuting developed models.

1.3.1 Thesis Objectives

To summarize the problem description, this thesis aims to achieve the following:

- The identification and systematization of phenomena, both mechanical and electromagnetic, that take precedence in high-speed machines.

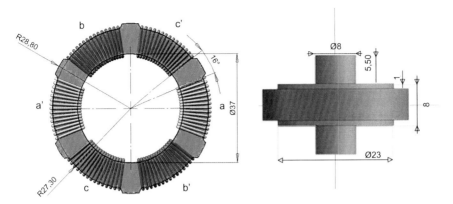

Fig. 1.3 The stator and the rotor with the dimensions in mm

- Define the speed limits of permanent magnet machines and correlating those limits with basic parameters of the machines.
- Electromagnetic, structural (elastic) and rotordynamical modeling of a high-speed permanent magnet machine.
- The design and realization of a low-stiffness high-speed spindle drive.

1.3.2 Test Setup: A High-Speed PM Motor in Aerostatic Bearings

The work on the high-speed spindle drive resulted in a practical setup. Since it is used throughout this thesis for validation of the models and concepts, a very short preview of the setup is given here. Details on the design of the setup are presented in Chap. 7.

The setup consists of a radial-flux slotless PM motor with a disc-shaped rotor which is supported by aerostatic bearings. A drawing of the stator and the rotor is given in Fig. 1.3. A magnet is applied onto the rotor in a ring shape and magnetized diametrically. The stator is wound toroidally and fitted into the bearings' housing.

In the thesis the setup will be referred to as the *test setup* and the PM machine as the *test machine* or the *test motor*.

1.4 Thesis Outline

Chapter 2—High-Speed PM Machines: Applications, Trends and Limits

In the chapter, an overview of current and prospective applications of permanent magnet machines is presented and speed limits of those machines are studied with particular focus on small-size, high-speed applications. Several prominent and promising applications of very high-speed machines are reported. The term *high-speed* is discussed and its meaning in this thesis is specified. The results of an empirical survey on the correlation between rated powers and speeds of existing high-speed machines are presented. Physical factors that shape the speed limits of PM machines in general are defined. The last section of the chapter attempts to theoretically correlate the maximum rated powers and speeds of PM machines with respect to their inherent limits, as defined in the preceding sections.

Chapter 3—Electromagnetic Modeling of Slotless PM Machines

The chapter presents magnetostatic modeling of a slotless PM machine that will form the basis of the electromagnetic optimization of the test machine. Based on the model of the magnetic field, other quantities of the machine—no-load voltage, torque, inductance, unbalanced magnetic force and losses—are derived. The main purpose of the chapter is to distinguish and model the most important electromagnetic parameters of high-speed PM machines. The modeling aims at suitable representation of dominant phenomena that can serve as a good basis for designing high-speed (slotless) PM machines.

Chapter 4—Structural Aspects of PM Rotors

Structural modeling and optimization of high-speed PM rotors are the focus of this chapter. The aim of the chapter is to model the influence of rotational speed and mechanical fittings on stress in a high-speed rotor, while also considering the operating temperature. Through analytical modeling, structural limits for the rotor speed are determined and quantified. At the same time, an analytical representation of the limiting parameters implies a relatively simple approach for the optimization of the rotor structure. This optimization approach will form the basis for the design of a carbon-fiber sleeve for the rotor of the test machine.

Chapter 5—Rotordynamical Aspects of High-Speed Electrical Machines

The chapter gives qualitative insight into important dynamical aspects of high-speed rotors through analytical modeling. The goal of the chapter is to define the dynamical limits for the rotor speed and to correlate those limits with machine parameters. A theoretical study on rotation stability gives an assessment of the threshold speed of unstable, self-excited vibrations. Based on this, practical speed limits of electrical machines are defined with respect to the critical speeds of the rotor-bearings system and bearing properties. Basic correlations between critical speeds and parameters of the rotor-bearing system are modeled. Lastly, the unbalance response of rigid rotors is analytically modeled and the suitability of different rotor geometries for high-speed rotation is analyzed. The conclusions drawn from this chapter strongly influenced the final spindle concept.

Chapter 6—Bearings for High-Speed Machines

The goal of the chapter is to study different types of bearings with respect to their suitability for high-speed rotation. The chapter is primarily concerned with bearings that have been the most promising for high-speed rotation: (hybrid) ball bearings, externally pressurized (or static) air bearings and magnetic bearings. A general overview and comparison of these bearing types is given.

Chapter 7—Design of the High-Speed-Spindle Motor

The chapter presents the design of the spindle (test) motor, from conceptual design to the electromagnetic and structural optimization of the motor. The analyses and models presented in Chaps. 2–6 determined the spindle-motor design: the definition of speed limits of PM machines greatly affected the conceptual design of the new spindle drive while the developed models formed an analytical basis for the motor design and optimization. In the first part of the chapter, the development of new spindle concepts within the Microfactory project group at TU Delft is presented. Geometric and electromagnetic design of the spindle motor is explained and then evaluated using FEM. Lastly, the optimization of the rotor retaining sleeve is presented and the method for rotor production is defined.

Chapter 8—Control of the Synchronous PM Motor

The chapter presents a new control method developed for control of the high-speed motor. For stable, sensorless control of a high-speed PM synchronous motor an open-loop, I/f control method is proposed in which stabilizing control from the standard V/f approach has been incorporated. The controller was successfully implemented in a standard control-purpose DSP and tested in the practical setup.

Chapter 9—Experimental Results

The chapter gives a description of the test setup and experiments which were carried out to gather practical data which further allowed the assessment of the developed models and chosen design approach. The main tests performed for verification of the models developed in the thesis are the speed-decay and locked-rotor tests. The overall performance of the electric drive during operation is also reported and discussed in the chapter.

Chapter 10—Conclusions and Recommendations

In the last thesis chapter, theoretical and practical work presented in the thesis is revisited and evaluated. Important conclusions are drawn with respect to the validation of the developed models, speed limits of PM machines and performance and applicability of the spindle-drive design. The chapter lists the main contributions of the thesis and gives recommendations for future work.

References

1. J. Chae, S. Park, T. Freiheit, Investigation of micro-cutting operations. Int. J. Mach. Tools Manuf. **46**(3–4), 313–332 (2006)
2. D. Bourell, K.F. Ehmann, M.L. Culpepper, T.J. Hodgson, T.R. Kurfess, M. Madou, K. Rajurkar, R.E. DeVor, *International Assessment of Research and Development in Micromanufacturing* (World Technology Evaluation Center, Baltimore, 2005)
3. H. Wicht, J. Bouchaud, *NEXUS Market Analysis for MEMS and Microsystems III* 2005–2009, Setting the Pace for Micro Assembly Solutions, 2005
4. X. Luo, K. Cheng, D. Webb, F. Wardle, Design of ultraprecision machine tools with applications to manufacture of miniature and micro components. J. Mater. Process. Technol. **167**(2–3), 515–528 (2005)
5. T. Masuzawa, State of the art of micromachining. CIRP Ann. Manuf. Technol. **49**(2), 473–488 (2000)
6. G.L. Benavides, D.P. Adams, P. Yang, *Meso-Machining Capabilities*, Sandia Report, 2001
7. D. Huo, K. Cheng, F. Wardle, Design of a five-axis ultra-precision micro-milling machineâŁ"UltraMill. Part 1: holistic design approach, design considerations and specifications. Int. J. Adv. Manuf. Technol. **47**(9–12), 867–877 (2009)
8. Y. Okazaki, N. Mishima, K. Ashida, Microfactory-concept, history, and developments. J. Manuf. Sci. Eng. **126**, 837 (2004)
9. Y. Bang, K. Lee, S. Oh, 5-axis micro milling machine for machining micro parts. Int. J. Adv. Manuf. Technol. **25**(9), 888–894 (2005)
10. M. Vogler, X. Liu, S. Kapoor, R. DeVor, K. Ehmann, Development of meso-scale machine tool (mMT) systems. Trans. NAMRI/SME **30**, 653–661 (2002)
11. E. Kussul, T. Baidyk, L. Ruiz-Huerta, A. Caballero-Ruiz, G. Velasco, L. Kasatkina, Development of micromachine tool prototypes for microfactories. J. Micromech. Microeng. **12**, 795 (2002)
12. S.W. Lee, R. Mayor, J. Ni, Dynamic analysis of a mesoscale machine tool. J. Manuf. Sci. Eng. **128**, 194 (2006)
13. H. Li, X. Lai, C. Li, Z. Lin, J. Miao, J. Ni, Development of meso-scale milling machine tool and its performance analysis. Front. Mech. Eng. China **3**(1), 59–65 (2008)
14. Y. Takeuchi, Y. Sakaida, K. Sawada, T. Sata, Development of a 5-axis control ultraprecision milling machine for micromachining based on non-friction servomechanisms. CIRP Ann. Manuf. Technol. **49**(1), 295–298 (2000)
15. J.-K. Park, S.-K. Ro, B.-S. Kim, J.-H. Kyung, W.-C. Shin, J.-S. Choi, A precision meso scale machine tools with air bearings for microfactory, in *5th International Workshop on Microfactories, Besancon, France*, 2006
16. Nanowave, Nano Corporation (2006), http://www.nanowave.co.jp/index_e.html
17. Robonano α-0iA Brochure, Fanuc Ltd., http://www.fanuc.co.jp/en/product/robonano/index.htm
18. R. Blom, M. Kimman, H. Langen, P. van den Hof, R.M. Schmidt, Effect of miniaturization of magnetic bearing spindles for micro-milling on actuation and sensing bandwidths, in

Proceedings of the Euspen International Conference, EUSPEN 2008, Zurich, Switzerland, 2008

19. D. Dornfeld, S. Min, Y. Takeuchi, Recent advances in mechanical micromachining. CIRP Ann. Manuf. Technol. **55**(2), 745–768 (2006)
20. A.G. Phillip, S.G. Kapoor, R.E. DeVor, A new acceleration-based methodology for micro/meso-scale machine tool performance evaluation. Int. J. Mach. Tools Manuf. **46**(12–13), 1435–1444 (2006)
21. T. Schaller, L. Bohn, J. Mayer, K. Schubert, Microstructure grooves with a width of less than $50\,\mu m$ cut with ground hard metal micro end mills. Prec. Eng. **23**, 229–235 (1999)
22. C.R. Knospe, Active magnetic bearings for machining applications. Control Eng. Pract. **15**(3), 307–313 (2007)
23. C.W. Lin, J.F. Tu, J. Kamman, An integrated thermo-mechanical-dynamic model to characterize motorized machine tool spindles during very high speed rotation. Int. J. Mach. Tools Manuf. **43**(10), 1035–1050 (2003)
24. A. Binder, T. Schneider, High-speed inverter-fed ac drives, in *Electrical Machines and Power Electronics, 2007. ACEMP'07. International Aegean Conference on*, pp. 9–16, 10–12 Sept 2007
25. A. Borisavljevic, H. Polinder, J. Ferreira, On the speed limits of permanent-magnet machines. IEEE Trans. Ind. Electron. **57**(1), 220–227 (2010)
26. H. Langen, Microfactory research topics in the Netherlands, in *The 5th International Workshop on Microfactories, Besancon, France*, Oct 2006
27. J.F. Gieras, High speed machines, in *Advancements in Electric Machines (Power Systems)*, ed. by J.F. Gieras (Springer, Berlin 2008)
28. High Speed, Calnetix Inc., http://www.calnetix.com/highspeed.cfm
29. L. Zhao, C. Ham, L. Zheng, T. Wu, K. Sundaram, J. Kapat, L. Chow, C. Siemens, A highly efficient 200,000 rpm permanent magnet motor system. IEEE Trans. Magn. **43**(6), 2528–2530 (2007)
30. C. Zwyssig, J.W. Kolar, S.D. Round, Megaspeed drive systems: pushing beyond 1 million r/min. IEEE/ASME Trans. Mechatron. **14**(5), 598–605 (2009)
31. P. Mellor, M. Al-Taee, K. Binns, Open loop stability characteristics of synchronous drive incorporating high field permanent magnet motor. IEE Proc. B Electr. Power Appl. **138**(4), 175–184 (1991)
32. M. Kimman, H. Langen, J. van Eijk, R. Schmidt, Design and realization of a miniature spindle test setup with active magnetic bearings, in *Advanced Intelligent Mechatronics, 2007 IEEE/ASME International Conference on*, pp. 1–6, 4–7 Sept 2007

Chapter 2
High-Speed PM Machines: Applications, Trends and Limits

2.1 Introduction

High-speed permanent magnet machines are the focus of this thesis; this chapter offers an overview of their current and prospective applications and a theoretical study of their limits.

The following section explains the reasons behind the increased use of PM machines with particular attention to small-size, high-speed applications. Several prominent and promising applications of very high-speed machines are reported. In Sect. 2.3 the term high-speed is discussed and its meaning in this thesis is specified. Sect. 2.4 presents results of an empirical survey of the correlation between rated powers and speeds of existing high-speed machines; in particular, the section points out trends of speed increase of slotted and slotless PM machines.

The rest of the chapter is dedicated to speed limits of PM machines. Physical factors that lie behind speed limits of PM machines in general are defined in Sect. 2.5. Finally, Sect. 2.6 attempts to theoretically correlate maximum rated powers and speeds of PM machines with respect to their inherent limits defined in the preceding section.

2.2 PM Machines: Overview

Permanent magnet machines have become increasingly popular in the last two decades. They have replaced induction machines in a great number of converter-fed electrical drives and motion control systems[1] [1, 2]. Besides, many new applications, which require high-performance electrical machines, are invariably

---

[1] A simple indicator of the prevalence of PM machines can be achieved by examining the increase of the relative number of associated academic papers. For illustration, the ratio between number of papers that is associated with PM/PM synchronous/brushless DC and induction/asynchronous motors in the IEEE internet base grew from 0.13 for the period before 1990 to 0.54 after 1990 and to 0.65 after year 2000.

A. Borisavljević, *Limits, Modeling and Design of High-Speed Permanent Magnet Machines*,   13
Springer Theses, DOI: 10.1007/978-3-642-33457-3_2,
© Springer-Verlag Berlin Heidelberg 2013

linked to PM machines. Two factors may be singled out as crucial for such development.

Cost competitiveness and availability of rare-earth magnets. After years of rather slow advance of SmCo magnets on the market, in the eighties, rapid commercialization of NdFeB magnets took place [3]. In the beginning of the nineties the production of rare-earth magnets started booming in China, relying on its large deposits of rare earths and cheap mining [3]. The price of rare-earth magnets went down which removed the main obstacle for using strong magnets in a wide range of applications; accordingly, PM machines became cost-competitive [4, 5]. However, not only has the price of the magnet been reduced in the past 20 years, a palette of different high-energy magnets has become readily available: sintered and plastic-bonded magnets with different energy products, Curie's temperatures, corrosion resistance, shape flexibility, etc [6, 7]. This has allowed the use of PM machines by even highly demanding customers such as the aircraft industry or military.

Emergence of efficiency-driven applications. Importance of economical exploitation of resources (fuel in particular) has brought about need for compact and efficient electromechanical systems. In practice it means that more-efficient electrical machines are to either replace traditionally pneumatic, hydraulic and combustion engines (e.g. in aircraft sub-systems, electric cars) or become physically integrated and directly coupled with mechanical systems (e.g. machining spindles, flywheels, turbines).

Qualities of PM machines make them a preferred choice for such systems. Owing to strong magnets, PM machines can have high power densities and also achieve high efficiencies due to no excitation losses, no magnetizing currents and very low rotor losses.

Virtue of high power density becomes particularly important for applications where low volume/weight and high-speeds are important. Undoubtedly, PM machines dominate the field of small high-speed machines [8]. This can be ascribed to their magnetic excitation—air-gap flux density of PM machines is determined mainly by the quality of utilized permanent magnets and does not depend on the size of the machine. Current-excited machines, on the other hand, lack space for conductors in small volumes and thus have comparatively smaller power densities.

For the same power requirement, a high-frequency design reduces size and weight of an electrical machine. High-frequency also often means elimination of power transmission elements. Thus, with downsizing and integration the resulting machinery becomes more efficient, lighter and even portable [9]. In the rest of this section a few typical applications for high-speed PM machines are discussed.

Miniature gas turbines are an exemplary application for high-speed generators. Small gas turbines are a promising means of converting fuel energy into electricity. Unlike large turbines whose output power is transmitted to generators via gears, these small gas turbines are conceived to be directly coupled with high-speed generators [9]. Resulting device would be a highly efficient power unit suitable as a portable power supply or a part of a distributed power network. Reliable gas turbines that utilize PM generators are available on the market [10]. Still, there is a great interest in academia for developing reliable high-speed generators that would keep pace with newly developed turbines which are capable of rotating at speeds beyond 1 million rpm [11, 12]. A good example of such an effort in academia is work on high-speed generators in ETH Zurich [13].

Waste heat from a gas turbine may be further used to heat water or space. As stated in [14]: "Because electricity is more readily transported than heat, generation of heat close to the location of the heat load will usually make more sense than generation of heat close to the electrical load." Such an idea lies behind concept of combined heat and power (CHP) [9, 15] systems which, according to current predictions, will be employed as highly efficient heating, cooling and power systems of buildings in coming years [16].

Machining spindles were already thoroughly discussed in Introduction: integration of a PM motor with a spindle gives way to efficiency, compactness and high speeds for production of complex 3D parts. Spindles for medical tools represent also an attractive application for small high-speed machines. Today, dental drilling spindles are mostly driven by air turbines, however, replacing the turbines with electrical drives would facilitate adjustable speed and torque of the spindles and reduce number of hand-pieces needed by a dentist [17].

The new trend of replacing mixed secondary power systems in aircraft with electrical ones has brought forth requirements for light-weight, fault-tolerant machines [18]. To achieve powerful engines in small volumes high-speed machines are necessary.

Integrated with flywheel a high-speed PM machine forms an electro-mechanical (EM) battery which is efficient and long-lasting device for energy storage [19, 20]. These EM batteries have been used in many applications, such as hybrid cars, locomotives [8] and spacecraft [21].

2.3 Defining *High-Speed*

At this point, it would be wise to define the speed which will be taken as the decisive factor in this thesis when naming a machine a high-speed machine. As long as technology permits, arbitrary high rotational speeds could be simply achieved by scaling down the machine. Therefore, it would not be sensible to take rotational speed as a sole criterion for the high speed.

The tangential speed at the outer rotor radius is often taken as a criterion when defining high-speed because it also takes into account the size of a machine. Such reasoning could make sense since one of major limiting factors for the rotational speed, mechanical stress in the rotor, is dependent on the tangential rotor speed. However, this criterion practically represents the degree of machine's mechanical utilization and it would favor very large generators that operate at 50/60 Hz which are hardly perceived as high-speed machines [8]. For that reason, as Binder and Schneider [8] point out, only inverter-fed or variable-speed machines can, in common understanding, be called high-speed.

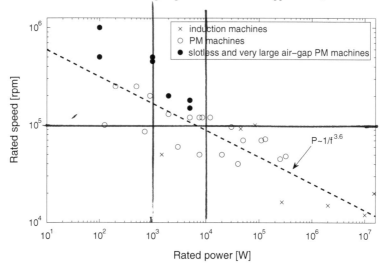

Fig. 2.1 Diagram of rated speeds and powers of existing high-speed machines, [10, 13, 17, 20, 21, 25–48] © 2010 IEEE

In [22] super high-speed machines are classified according to operating power and rotational speed. A numerical limit proposed in [23] correlating power limit with the rotational speed of an electrical machine was used as a criterion for super high-speed in [22]. This type of relationship has commonly been used to evaluate operating speed range of machines [13, 24]. Binder and Schneider [8] also empirically found a correlation between rated powers and speeds of super-high-speed machines.

Consequently, this thesis will focus on variable-speed PM machines of small and medium size that have high speed with respect to their power. This correlation will be discussed more in the rest of this chapter. Rated power of the machines of interest is typically below 500 kW—the power range where PM machines appear to be prevalent. Nevertheless, a good part of the analyses found in the thesis is applicable to a broad range of high-speed electrical machines.

2.4 Survey of High-Speed Machines

Based on collected data on commercially available high-speed machines and machines developed in academia [10, 13, 17, 20, 21, 25–48], a diagram of rated powers and speeds of high-speed permanent magnet and induction machines was made.[2]

Relation between rated powers and speeds in the diagram is in a good agreement with the correlation empirically found by Binder and Schneider [8]. Namely, from

---

[2] Most of the content of this section has been taken from Borisavljevic et al. [49], © 2010 IEEE.

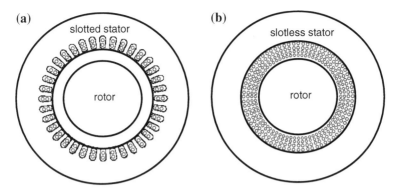

Fig. 2.2 Cross-section of a **a** slotted machine with a large air gap and **b** slotless machine © 2009 IEEE

a study on published data on high-speed AC machines, the authors obtained the relationship: $\log f = 4.27 - 0.275 \log P$ (consequently, $P \sim 1/f^{3.6}$), which is presented with a dashed line in the diagram.

The diagram (Fig. 2.1) shows prevalence of PM machines among small high-speed machines. To the author's knowledge, no machine other than permanent magnet has been reported to operate beyond speed of 100.000 rpm.

Slotless PM machines show a trend of the highest increases in speeds and extremely fast examples have been reported both on the market and in academia [17, 25]. Bianchi et al. [50] illustrated advantages of using a slotless, rather than a slotted, stator in high-speed PM machines. The authors optimized, constructed and, finally, assessed performance of machines of both types. Optimum flux density in very high-speed PM machines is usually low [50], affecting designs to result in machines with large effective air-gaps. Instead of increasing mechanical air-gap, it is sensible to replace stator teeth with conductors. In this manner, the increase of conductors' area enables rise of the rated current, and that, in turn, partly compensates loss of power density due to reduced air-gap flux density.

2.5 Speed Limits of PM Machines

Boost in speeds of PM machines is linked with overcoming or avoiding a number of machine limitations. Various physical parameters (stress, temperature, resonant frequencies) can limit the speed of an electrical machine. Aside from speed, these variables are also affected by power, size and machine electrical and magnetic loading.[3]

Only a small number of academic papers discuss the limits of high-speed machines. In the paper by Slemon [51] parameters of surface-mounted PM machines

---

[3] A good part of this section has been taken from Borisavljevic et al. [17], © 2010 IEEE.

were correlated with respect to physical and technological constraints. The paper derived general approximate expressions for acceleration and torque limits.

Bianchi et al. [50] evaluated thermal and PM-demagnetization limits of different types of high-speed PM machines. Those limits were taken into account in the proposed optimization procedure through constraints imposed on the design variables. In a later paper by the same authors [52] the demagnetization limit was disregarded in slotless PM machine design as it is too high to be reached.

In extremely high-speed machines, however, mechanical factors such as stress and vibrations, rather than electromagnetically induced heating, are likely to cause failure of the machine. A good modeling of elastic behavior and constraints of a rotor of an electrical machine can be found in the thesis of Larsonneur [53]. Other authors also took mechanical constraints into account when designing a PM rotor, e.g. [54, 55]. Finally, stability of rotation has been analyzed comprehensively in the field of rotordynamics [56–59], however some important conclusions on rotordynamical stability have not been included in literature on electrical machines.

Speed limitations of PM machines in general are the topic of this section. Only physical limits that are inherent to PM machines will be discussed; speed limitations associated with bearing types, power electronics, control or technological difficulties will not be considered. Defining and quantification of the machine speed limits in the rest of the thesis will have an essential impact on the design of the test machine.

The limit that is common to all machines is the thermal limit. The thermal behavior of a machine depends on power losses that are further dependent on current and magnetic loading, as well as rotational speed. These parameters will be correlated with machine size and rated power in the next section in order to see how the speed is limited.

The strength or stress limit is inherently connected with the elastic properties of the materials used in the rotor. Mechanically, a PM rotor can be modeled as a compound of two or three cylinders: a rotor magnet that either leans on a rotor iron shaft or is a full cylinder, and an enclosure of the magnet. Given the interference fits between the adjacent cylinders, maximum tangential speed $v_{t,max}$ of the rotor can be determined (see Chap. 4) at which either maximum permissible stress is reached in one of the cylinders or contact at boundary surface(s) is lost (Fig. 2.2).

The third limit that will be considered is related to rotordynamical properties of the mechanical system. For every rotor-bearing system two types of vibrations can occur that can (or will) limit the rotational speed of the rotor: resonant and self-excited. Resonant vibrations occur when the speed of the rotor coincides with one of the resonant frequencies. In literature are those rotor speeds referred to as *critical speeds*, among which those connected with flexural modes in the rotor are particularly dangerous. Self-excited vibrations make rotation impossible, that is, unstable, and they commence after a certain threshold speed [60].

Tackling the problem of rotor vibrations is a very complicated task that includes the mechanical design of the rotor and design of the bearings. However, the speed range in which the rotor will operate is usually decided on in the initial design stage: the working speed range must be sufficiently removed from the resonant speeds and below the threshold of instability. For a given rotor geometrical profile and bearings'

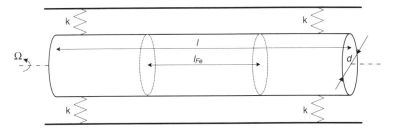

Fig. 2.3 Rotor-bearings configuration © 2010 IEEE

stiffness the maximum ratio between the rotor length and diameter can be defined, Fig. 2.3:

$$\lambda_{\max} = \frac{l}{d} = \frac{l_{Fe}}{\alpha 2 r_r}. \tag{2.1}$$

so as to insure stable rotation at the target maximum rotor speed.

From (2.1) maximum ratio between active rotor length (active machine length) and the rotor radius is determined:

$$\lambda_{Fe,\max} = \frac{l_{Fe}}{2 r_r} = \alpha \lambda_{\max}, \tag{2.2}$$

where $\alpha = l_{Fe}/l$ is ratio between active and total rotor length.

Parameter $\lambda_{\max}$ can be viewed as a figure of the rotor slenderness or, conversely, robustness, with lower values for a more robust rotor.

2.6 Limits and Rated Machine Parameters

The goal of this section is to theoretically correlate rated power, speed and size of PM machines and, as a result, account for speed-power relationship in the diagram in Sect. 2.4. The correlation will be drawn considering the limiting factors mentioned in the previous section. The maximum slenderness $\lambda_{\max}$ and tangential speed $v_{t,max}$ will be assumed known, as defined by mechanical properties of the system. Temperature in a machine depends on losses and on the cooling capability thus the surface loss power density will be maintained constant. The derivation is similar to the one in [8], with focus on loss influence in different types of PM machines (See footnote 3).

To analyze interdependence of power, size and speed the following approximate equation for output power will be used:

$$P_{nom} = \Omega T = \underbrace{2\pi f_{nom}}_{\Omega} \cdot \underbrace{F_d \cdot 2\pi r_s l_{Fe} \cdot r_s}_{F} \approx 2\pi^2 B_g A_c r_s^2 l_{Fe} f_{nom}. \tag{2.3}$$

In the equation P_{nom}, T and f_{nom} are, respectively, rated power, torque and frequency and $F_d \approx B_g A_c / 2$ [61] denotes machine's maximum force density, where B_g is maximum air-gap flux density and:

$$A_c \approx \begin{cases} k_w k_{fill} J h_s \frac{b_s}{b_s + b_t}, & \text{for a slotted machine} \\ k_w k_{fill} J l_w, & \text{for a slotless machine} \end{cases} \tag{2.4}$$

is current loading. In (2.4) J is amplitude of sinusoidal stator current density, h_s, b_s and b_t are slot height, slot width and tooth width of a slotted machine, l_w is conductors' area thickness of a slotless machine and k_w and k_{fill} are winding and fill factors, respectively. If the ratio of slot to tooth width is kept constant and having $h_s, l_w \sim r_r$, the expressions (2.4) yield:

$$A_c \sim J r_r. \tag{2.5}$$

Using (2.1) in Eq. (2.3) and neglecting the air-gap length ($r_s \approx r_r$) we obtain:

$$P_{nom} = 4\pi^2 B_g A_c \alpha \lambda_{max} r_r^3 f_{nom}. \tag{2.6}$$

Due to increased losses, to maintain a given cooling capability, magnetic and current loading must be lowered if frequency is increased. Consequently, the ratio between maximum permissible power dissipation and cooling surface area is maintained constant:

$$\frac{P_{diss}}{S} = const, \tag{2.7}$$

while keeping in mind that $S \sim r_r^2$.

An analysis of the influence of the machine size and frequency on dominant, frequency-dependent losses will follow.

At very high frequencies iron losses in the stator are estimated as:

$$P_{Fe} = C \cdot m_{Fe} f_e^2 B_g^2 \sim r_r^3 f_{nom}^2 B_g^2, \tag{2.8}$$

where f_e:

$$f_e = \frac{N_{poles}}{2} f_{nom} \tag{2.9}$$

is electrical frequency and m_{Fe} is mass of the iron core.

Copper losses can be divided into two parts: (i) the conduction loss part $P_{Cu,skin}$, that also includes a rise in loss caused by the reduction of effective conductors cross-section due to skin-effect; and (ii) the proximity loss part $P_{Cu,prox}$, that accounts for eddy-current loss in the conductors due to the pulsating magnetic fields from the rotor magnet and neighboring conductors:

$$P_{Cu} = P_{Cu,skin} + P_{Cu,prox}. \tag{2.10}$$

According to [62] we can estimate the skin-effect conduction loss as:

$$P_{Cu,skin} = F(\phi) \cdot P_{Cu,DC} = \underbrace{(F(\phi) - 1) I^2 R_{DC}}_{\text{skin-effect}} + \underbrace{I^2 R_{DC}}_{\text{DC}}, \tag{2.11}$$

where ϕ is the ratio between conductor diameter and the skin-depth:

$$\phi = \frac{d_{Cu}}{\delta_{skin}\sqrt{2}} = d_{Cu}\sqrt{\frac{\pi \sigma_{Cu} \mu_0 f_e}{2}}, \tag{2.12}$$

and function $F(\phi)$ is defined as:

$$F(\phi) = \frac{\phi}{2} \cdot \frac{(ber(\phi)bei'(\phi) - bei(\phi)ber'(\phi))}{ber'^2(\phi) + bei'^2(\phi)}. \tag{2.13}$$

For reasonable conductor diameters and electrical frequencies of high-speed machines the skin-effect part of Eq. (2.11) comprises less than one percent of the total value of $P_{Cu,skin}$—see Fig. 3.17 in Sect. 3.6.2—and can thus be neglected. Therefore:

$$P_{Cu,skin} \approx P_{DC} = I^2 R_{DC} = J^2 S_w \rho_{Cu} l_{Cu} \sim J^2 r_r^2 l_{Fe} \sim J^2 r_r^3. \tag{2.14}$$

In Eq. (2.14) S_w is the copper cross-section area of the windings and ρ_{Cu} is the copper resistivity.

Regarding the proximity-effect part of the copper losses, in the example of slotless PM machines, it will primarily be influenced by the rotor permanent magnet field rather than by the field of the neighboring conductors. This part of losses can be estimated as [63] (for detail see Sect. 3.6.2):

$$P_{Cu,prox,slotless} = \frac{B_g^2 (2\pi f_{nom})^2 d_{Cu}^2}{32\rho_{Cu}} V_{Cu} \sim r_r^3 B_g^2 f_{nom}^2. \tag{2.15}$$

where V_{Cu} is volume of the conductors (copper).

In a slotted machine the influence of neighboring conductors on proximity loss is dominant, but dependence of this loss on frequency takes on a more complicated form. The loss in a single conductor can be expressed in a following form (see [62]):

$$P_{Cu,prox}^1 = \frac{G(\phi)}{\sigma_{Cu}} \cdot H_{e,rms}^2 \cdot l_{Fe} \tag{2.16}$$

where $H_{e,rms}$ is the rms value of magnetic field strength of the neighboring conductors and:

$$G(\phi) = 2\pi\phi \cdot \frac{(ber_2(\phi)ber'(\phi) - bei_2(\phi)bei'(\phi))}{ber^2(\phi) + bei^2(\phi)}. \tag{2.17}$$

For a constant conductor diameter dependence of the expression $G\left(\phi\right)$ on frequency is very closely quadratic—see Fig. 3.18 and Eq. (3.105) in Sect. 3.6.2—and the total loss caused by proximity effect for the slotted machine case can be approximated as:

$$P_{Cu,prox,slotted} \sim J^2 f_{nom}^2 V_{Cu} \sim J^2 f_{nom}^2 r_r^3 \tag{2.18}$$

To maintain the permissible surface loss density (2.7) in a slotless machine, from (2.8), (2.15) and (2.14) restrictions for magnetic flux density and current density are obtained as:

$$B_g \sim 1/(\sqrt{r_r}\, f_{nom}) \tag{2.19}$$

and

$$J \sim 1/\sqrt{r_r}. \tag{2.20}$$

Finally from (2.5), (2.6), (2.19) and (2.20) the correlation between the rated power and size for slotless PM machines is obtained:

$$P_{nom} \sim r_r^3. \tag{2.21}$$

In the slotted machine case, expressions for DC- (2.14) and frequency-dependent copper loss (2.18) yield different restrictions for current density: $J \sim 1/\sqrt{r_r}$ and $J \sim 1/(\sqrt{r_r}\, f_{nom})$. Correlation between rated parameters of slotted PM machines can be expressed in a more general form as:

$$P_{nom} \sim \frac{r_r^3}{f_{nom}^k}, \quad 0 < k < 1, \tag{2.22}$$

where k depends on portion of frequency-dependent losses in total copper loss.

The last two equations explain trend of higher increase in rated speeds of slotless PM machines with respect to their slotted counterparts, as presented in Sect. 2.4.

Further, from (2.21) and (2.22) it is evident that it is not possible to gain power density by merely scaling down the machine and increasing its rated speed. Unfortunately, with increasing operating speed (to obtain the same machine power), losses in the machine will increase rapidly. To maintain a given cooling capability, magnetic and current loading must be lowered if frequency is increased.

However, the rationale behind using high-speed machines is compactness and efficiency of a particular electro-mechanical system and not of the machine itself.

Air-friction loss has not been analyzed yet in this section and it represents a great portion of overall losses in high-speed machines. Magnetic and current loading, however, do not influence air-friction and it is dependent solely on mechanical quantities: size, speed and roughness of the rotor and length of the air-gap. Power of the air-friction drag is given by:

$$P_{af} = k_1 C_f\left(v_t, l_g\right) \rho_{air} \pi \Omega^3 r_r^4 l. \tag{2.23}$$

Empirical estimations of the friction coefficient C_f for typical rotor geometries can be found in the thesis of Saari [64]. If the air-friction power is divided with the rotor surface area $2\pi r_r l$ a surface density of air-friction loss power is obtained:

$$q_{af} = \frac{1}{2} k_1 C_f \left(v_t, l_g\right) \rho_{air} v_t^3. \tag{2.24}$$

Temperature in the air-gap is proportional to this power density. Whether this temperature will jeopardize operation of a machine depends on the coefficient of thermal convection on the rotor surface and that coefficient is rather difficult to assess. In any case, tangential rotor speed is a potential limiting figure for machine with respect to air-friction loss.

If tangential speed $v_t = 2\pi r_r f_{nom}$ is used in (2.21) and (2.22) instead of the radius, the following correlations are obtained:

$$P_{nom} \sim \begin{cases} v_t^3 / f_{nom}^{3+k}, & \text{for a slotted machine} \\ v_t^3 / f_{nom}^3, & \text{for a slotless machine} \end{cases} \tag{2.25}$$

Equation (2.25) practically establish connection between the nominal powers of PM machines, their rotor radii (sizes) and nominal rotational frequencies. Usually, tangential speed in PM machines is limited by stress in the rotor and today does not exceed 250 m/s (the highest reported rotor tangential speed is 245 m/s and it is associated with a very large turbine generator in Germany, operating at 50 Hz [8]). Hence, correlation between nominal powers of PM machines with their *highest permissible* nominal speeds can be expressed in a general form as:

$$P_{nom} \sim \frac{1}{f_{nom}^{3+k}}, \quad 0 < k < 1, \tag{2.26}$$

which Binder and Schneider [8] found with an approximate value of $k = 0.6$ for the frequency exponent (Fig. 2.1), however, by including also induction and homo-polar machines in the study.

2.7 Conclusions

The chapter gives an introduction to high-speed permanent magnet machines: it analyzes prominent and prospective applications of these machines, defines the attribute "high-speed" for the purpose of this thesis and presents a general theoretical study of speed limits of PM machines. By doing so, the chapter explains the reasoning behind choosing the slotless PM machine as the type of the test-motor and, more importantly, it defines the scope and outlook for the modeling and design presented in the thesis.

Two reasons are singled out for the prevalence of PM machines in the last two decades: (i) rapid commercialization and availability of rare-earth magnets and (ii) the emergence of efficiency-driven applications. Qualities of PM machines—high efficiency and power density—have made them a preferred choice for high-performance applications, particularly at medium and low powers.

High power density of PM machines can be ascribed to their magnetic excitation—air-gap flux density of PM machines is determined mainly by the quality of utilized permanent magnets and does not depend on the size of the machine. At the same time, relatively simple, cylindrical rotor can be robust enough to sustain the stress caused by the centrifugal force at high speeds.

In the thesis, *high-speed* machines are defined as variable-speed machines of small and medium size (typically, below 500 kW) that have high speed with respect to their power. Empirical study on such machines developed in industry and academia also showed the predominance of PM machines. Obtained relation between rated powers and speeds is in good agreement with the correlation empirically found by Binder and Schneider [8].

Slotless PM machines show a trend of the highest increase in speeds and extremely fast examples have been reported both on the market and in academia. Since the optimal air-gap flux density in high-speed PM machines for highest efficiency is usually low it is sensible to replace stator teeth with conductors and, in that way, partly compensates loss of power density due to reduced air-gap flux density.

Finally, the chapter identifies inherent (physical) speed limits of PM machines—thermal, structural (elastic) and rotordynamical. Taking into account their physical limits, the chapter theoretically correlates speed, power and size of PM machines and accounts for the correlations appearing from the empirical survey. It is concluded that, if the cooling method is maintained, it is not possible to gain power density by merely scaling down the machine and increasing its rated speed. This study can be viewed as the main contribution of this chapter.

References

1. Z. Zhu, K. Ng, D. Howe, Design and analysis of high-speed brushless permanent magnet motors, in *Electrical Machines and Drive, 1997 Eighth International Conference on (Conf. Publ. No. 444)*, pp. 381–385, 1–3 Sept 1997
2. K. Binns, D. Shimmin, Relationship between rated torque and size of permanent magnet machines. IEE Proc. Electr. Power Appl. **143**(6), 417–422 (1996)
3. S. Trout, Rare earth magnet industry in the USA: current status and future trends, in *17th International Workshop on Rare Earth Magnets and Their Applications*, University of Delaware, 2002
4. T. Shimoda, A prospective observation of bonded rare-earth magnets. IEEE Trans. J. Magn. Jpn. **8**(10), 701–710 (1993)
5. P. Campbell, Magnet price performance, in *Magnetic Business and Technology*, 2007
6. W. Rodewald, M. Katter, Properties and applications of high performance magnets, in *18th International Workshop on High Performance Magnets and Their Applications*, pp. 52–63, 2004

7. W. Pan, W. Li, L.Y. Cui, X.M. Li, Z.H. Guo, Rare earth magnets resisting eddy currents. IEEE Trans. Magn. **35**(5, Part 2), 3343–3345 (1999)
8. A. Binder, T. Schneider, High-speed inverter-fed ac drives, in *Electrical Machines and Power Electronics, 2007. ACEMP'07. International Aegean Conference on*, pp. 9–16, 10–12 Sept 2007
9. J.F. Gieras, High speed machines, in *Advancements in Electric Machines (Power Systems)*, ed. by J.F. Gieras (Springer, Berlin 2008)
10. E.A. Setiawan, Dynamics behavior of a 30Â kW capstone microturbine, in *Institut fuer Solare Energieversorgungstechnik eV (ISET)*, Kassel, Germany, 2007
11. K. Isomura, M. Murayama, H. Yamaguchi, N. Ijichi, N. Saji, O. Shiga, K. Takahashi, S. Tanaka, T. Genda, M. Esashi, Development of micro-turbo charger and micro-combustor as feasibility studies of three-dimensional gas turbine at micro-scale. ASME Conf. Proc. **2003**, 685–690 (2003)
12. J. Peirs, D. Reynaerts, F. Verplaetsen, Development of an axial microturbine for a portable gas turbine generator. J. Micromech. Microeng. **13**, S190–S195 (2003)
13. C. Zwyssig, S.D. Round, J.W. Kolar, An ultra-high-speed, low power electrical drive system. IEEE Trans. Ind. Electron. **55**(2), 577–585 (2008)
14. R. Lasseter, P. Paigi, Microgrid: a conceptual solution, in *Power Electronics Specialists Conference, 2004. PESC 04. 2004 IEEE 35th Annual*, vol. 6, pp. 4285–4290, 20–25 June 2004
15. P.A. Pilavachi, Mini-and micro-gas turbines for combined heat and power. Appl. Therm. Eng. **22**(18), 2003–2014 (2002)
16. MTT Recuperated Micro Gas Turbine for Micro CHP Systems, Micro Turbine Technology BV, http://www.mtt-eu.com/
17. C. Zwyssig, J.W. Kolar, S.D. Round, Megaspeed drive systems: pushing beyond 1 million r/min. IEEE/ASME Trans. Mechatron. **14**(5), 598–605 (2009)
18. G.J. Atkinson, *High Power Fault Tolerant Motors for Aerospace Applications*. Ph.D. Dissertation, University of Newcastle upon Tyne, 2007
19. S. Jang, S. Jeong, D. Ryu, S. Choi, Comparison of three types of pm brushless machines for an electro-mechanical battery. IEEE Trans. Magn. **36**(5), 3540–3543 (2000)
20. A. Nagorny, N. Dravid, R. Jansen, B. Kenny, Design aspects of a high speed permanent magnet synchronous motor/generator for flywheel applications, in *Electric Machines and Drives, 2005 IEEE International Conference on*, pp. 635–641, 15 May 2005
21. A.S. Nagorny, R.H. Jansen, D.M. Kankam, Experimental performance evaluation of a high-speed permanent magnet synchronous motor and drive for a flywheel application at different frequencies, in *Proceedings of 17th International Conference on Electrical Machines—ICEM*, 2006
22. M. Rahman, A. Chiba, T. Fukao, Super high speed electrical machines–summary, in *Power Engineering Society General Meeting, 2004. IEEE*, vol. 2, pp. 1272–1275, 10 June 2004
23. A. Maeda, H. Tomita, O. Miyashita, Power and speed limitations in high speed electrical machines, in *Proceedings of IPEC, Yokohama, Japan*, pp. 1321–1326, 1995
24. J. Oliver, M. Samotyj, R. Ferrier, Application of high-speed, high horsepower, ASD controlled induction motors to gas pipelines, in *5th European Conference on Power Electronics and Applications, EPE'93*, pp. 430–434, 1993
25. High Speed, Calnetix Inc., http://www.calnetix.com/highspeed.cfm
26. Danfoss Turbocor Compressors, Danfoss Turbocor, http://www.turbocor.com/
27. Ultra-High Speed Motors & Generators, SatCon Applied Technology, http://www.satcon.com/apptech/mm/uhs.php
28. Permanent Magnet Rotor with CFRP Rotor Sleeve, e+a Elektromaschinen und Antriebe AG, http://www.eunda.ch/
29. M. Caprio, V. Lelos, J. Herbst, J. Upshaw, Advanced induction motor endring design features for high speed applications, in *Electric Machines and Drives, 2005 IEEE International Conference on*, pp. 993–998, 15 May 2005
30. P. Beer, J.E. Tessaro, B. Eckels, P. Gaberson, High-speed motor design for gas compressor applications, in *Proceeding of 35th Turbomachinery Symposium*, pp. 103–112, 2006

31. M. Harris, A. Jones, E. Alexander, Miniature turbojet development at hamilton sundstrand the TJ-50, TJ-120 and TJ-30 turbojets, in *2nd AIAA "Unmanned Unlimited" Systems, Technologies, and Operations Aerospace, Land, and Sea Conference and Workshop & Exhibit*, pp. 15–18, 2003
32. The Smallest Drive System in the World, Faulhaber Group (2004), http://www.faulhaber.com/n223666/n.html
33. ATE Micro Drives, ATE Systems, http://www.ate-system.de/en/products/ate-micro-drives.html
34. U. Schroder, Development and application of high speed synchronous machines on active magnetic bearings, in *Proceedings of MAG'97, Industrial Conference and Exhibition on Magnetic Bearings, Alexandria, Virginia*, p. 79, August 1997
35. M. Ahrens, U. Bikle, R. Gottkehaskamp, H. Prenner, Electrical design of high-speed induction motors of up to 15 mW and 20000 rpm, in *Power Electronics, Machines and Drives, 2002. International Conference on (Conf. Publ. No. 487)*, pp. 381–386, 4–7 May 2002
36. M. Larsson, M. Johansson, L. Naslund, J. Hylander, Design and evaluation of high-speed induction machine, in *Electric Machines and Drives Conference, 2003. IEMDC'03. IEEE, International*, vol. 1, pp. 77–82, 1–4 Sept 2003
37. B.-H. Bae, S.-K. Sul, J.-H. Kwon, J.-S. Shin, Implementation of sensorless vector control for super-high speed pmsm of turbo-compressor, in *Industry Applications Conference, 2001. Thirty-Sixth IAS Annual Meeting. Conference Record of the 2001 IEEE*, vol. 2, pp. 1203–1209, 30 Sept 2001
38. M. Aoulkadi, A. Binder, G. Joksimovic, Additional losses in high-speed induction machineâŁ"removed rotor test, in *Power Electronics and Applications, 2005 European Conference on*, pp. 10 pp.–P.10, 0–0 2005
39. H.-W. Cho, S.-M. Jang, S.-K. Choi, A design approach to reduce rotor losses in high-speed permanent magnet machine for turbo-compressor. IEEE Trans. Magn. **42**(10), 3521–3523 (2006)
40. I. Takahashi, T. Koganezawa, G. Su, K. Ohyama, A super high speed pm motor drive system by a quasi-current sourceinverter. IEEE Trans. Ind. Appl. **30**(3), 683–690 (1994)
41. PCB Spindles, Westwind Air Bearings, http://www.westwind-airbearings.com/pcb/index.html
42. J. Oyama, T. Higuchi, T. Abe, K. Shigematsu, R. Moriguchi, The development of small size ultra-high speed drive system, in *Power Conversion Conference—Nagoya, 2007. PCC'07*, pp. 1571–1576, 2–5 Dec 2007
43. OilFree, Motorized Compressor Systems, Mohawk Innovative Technology, http://www.miti.cc/
44. J. Bumby, E. Spooner, J. Carter, H. Tennant, G. Mego, G. Dellora, W. Gstrein, H. Sutter, J. Wagner, Electrical machines for use in electrically assisted turbochargers, in *Power Electronics, Machines and Drives, 2004. (PEMD 2004). Second International Conference on (Conf. Publ. No. 498)*, vol. 1, pp. 344–349, 31 March 2004
45. BorgWarner Turbo & Emission Systems, http://www.turbos.bwauto.com
46. C. Zwyssig, M. Duerr, D. Hassler, J.W. Kolar, An ultra-high-speed, 500000 rpm, 1 kW electrical drive system, in *Proceedings of Power Conversion Conference—PCC'07*, pp. 1577–1583, 2007
47. A. Binder, T. Schneider, M. Klohr, Fixation of buried and surface-mounted magnets in high-speed permanent-magnet synchronous machines. IEEE Trans. Ind. Appl. **42**(4), 1031–1037 (2006)
48. O. Aglen, A. Andersson, Thermal analysis of a high-speed generator, *Industry Applications Conference, 2003. 38th IAS Annual Meeting. Conference Record of the*, vol. 1, pp. 547–554, 12–16 Oct 2003
49. A. Borisavljevic, H. Polinder, J. Ferreira, On the speed limits of permanent-magnet machines. IEEE Trans. Ind. Electron. **57**(1), 220–227 (2010)
50. N. Bianchi, S. Bolognani, F. Luise, Potentials and limits of high-speed pm motors. IEEE Trans. Ind. Appl. **40**(6), 1570–1578 (2004)

51. G. Slemon, On the design of high-performance surface-mounted pm motors. IEEE Trans. Ind. Appl. **30**(1), 134–140 (1994)
52. N. Bianchi, S. Bolognani, F. Luise, High speed drive using a slotless pm motor. IEEE Trans. Power Electron. **21**(4), 1083–1090 (2006)
53. R. Larsonneur, <Emphasis Type="Italic">Design and Control of Active Magnetic Bearing Systems for High Speed Rotation</Emphasis>. Ph.D. Dissertation, Swiss Federal Institute of Technology Zurich, 1990
54. C. Zwyssig, J.W. Kolar, Design considerations and experimental results of a 100 W, 500,000 rpm electrical generator. J. Micromech. Microeng. **16**(9), 297–307 (2006)
55. T. Wang, F. Wang, H. Bai, J. Xing, Optimization design of rotor structure for high speed permanent magnet machines, in *Electrical Machines and Systems, ICEMS. International Conference on*, pp. 1438–1442, 8–11 Oct 2007
56. T. Iwatsubo, Stability problems of rotor systems. Shock Vib. Inform. Digest (Shock and Vibration Information Center) **12**(7), 22–24 (1980)
57. J. Melanson, J.W. Zu, Free vibration and stability analysis of internally damped rotating shafts with general boundary conditions. J. Vib. Acoust. **120**, 776 (1998)
58. W. Kim, A. Argento, R. Scott, Forced vibration and dynamic stability of a rotating tapered composite timoshenko shaft: Bending motions in end-milling operations. J. Sound Vib. **246**(4), 583–600 (2001)
59. G. Genta, *Dynamics of Rotating Systems* (Springer, Berlin, 2005)
60. V. Kluyskens, B. Dehez, H. Ahmed, Dynamical electromechanical model for magnetic bearings. IEEE Trans. Magn. **43**(7), 3287–3292 (2007)
61. A. Grauers, P. Kasinathan, Force density limits in low-speed pm machines due to temperature and reactance. IEEE Trans. Energy Convers. **19**(3), 518–525 (2004)
62. J. Ferreira, Improved analytical modeling of conductive losses in magnetic components. IEEE Trans. Power Electron. **9**(1), 127–131 (1994)
63. E. Spooner, B. Chalmers, 'TORUS': a slotless, toroidal-stator, permanent-magnet generator. IEE Proc. B. Electr. Power Appl. **139**(6), 497–506 (1992)
64. J. Saari, *Thermal Analysis of High-Speed Induction Machines*. Ph.D. Dissertation, Acta Polytechnica Scandinavica, 1998

Chapter 3
Electromagnetic Modeling of Slotless PM Machines

3.1 Introduction

The adequate modeling of the electromagnetic (EM) field is the starting point for a design of an electrical machine. The models derived in this chapter will form the basis for the geometrical optimization of the test machine described in Chap. 7.

The applicability of slotless PM machines for very high-speed operation, which was illustrated in the previous chapter, motivated the choice of a slotless configuration for the test machine. Therefore, the modeling in this chapter will take into account only permanent magnet machines with slotless geometry.

An analytical approach to the machine modeling and design was adopted because only analytical expressions represent an efficient means for machine optimization. In addition, it is often stated that analytical models offer greater insight into the modeled phenomena than numerical methods. That is true, however, to a certain extent. Strictly applied, the modeling in this chapter leads to rather long and complex analytical expressions which hardly offer any insight into the influence of the underlying parameters. On the other hand, the reason for the seeming lack of given insight of the finite element (FE) method usually lies in limited knowledge of its users and not in (seemingly obscured) logic behind the used FE software.

Analytical EM modeling of slotless machines is not strictly confined to such machines and can be applied, with certain corrections, to a broader selection of PM machines. Namely, slotted machines are frequently modeled by transforming their geometry into an *equivalent* slotless geometry in which the resulting air gap length is somewhat extended using the Carter's factor which accounts for the fringing of the magnetic field around slots. However, such a model, naturally, cannot explain the effects of slotting on the magnetic field and correlated phenomena such as cogging torque and slot-harmonic induced losses.

The available literature on analytical modeling of electrical machines, including PM machines, is immense and a literature overview on the subject would be

A. Borisavljević, *Limits, Modeling and Design of High-Speed Permanent Magnet Machines*, 29
Springer Theses, DOI: 10.1007/978-3-642-33457-3_3,
© Springer-Verlag Berlin Heidelberg 2013

too broad to be of much interest or use to the reader. Furthermore, the model to be presented in this chapter is developed in a rather standard manner—starting from the Maxwell equations, the model reduces to Poisson's or Laplace's equations expressed over magnetic (vector) potential—which is generally known and used by electrical machines experts. Still, as a good reference for the topic one could point to the outstanding work of Zhu whose series of articles on analytical modeling of PM machines of different topologies [1–5] represents a comprehensive resource for the accurate modeling of PM machines. For the modeling in this chapter, though, the author primarily referred to the theses of Polinder [6] and Holm [7].

The EM model in this chapter is two-dimensional since a slotless machine with its large effective air gap cannot be accurately modeled without accounting for the curvature of the field in the air gap. Magnetic vector potential was chosen as the basic variable from which other field quantities are derived. The model is *magnetostatic* meaning that it neglects the reaction field of the eddy currents induced in the conductive parts of the machine. There are several reasons for this approach. Firstly, the influence of the reaction field of the (rotor) eddy currents on the total field and, consequently, other EM quantities, is normally very small (see Zhu's article [5]). Therefore, the magnetostatic model is quite sufficient for the machine design. Furthermore, a practical reason for the magnetostatic modeling is that the magnet of the test motor is of a plastic-bonded type which is chosen to maximally suppress eddy currents in the rotor. Finally, the power of eddy-current losses, particularly critical for the heating of the rotor magnet, can still be assessed based on the harmonics of the magnetostatic field [8, 9].

Although eddy currents are neglected when the magnetic field in the machine is modeled, their effect on losses in conductive machine parts is accounted for in Sect. 3.6.

Based on the model of the magnetic field in the machine, other quantities of the machine—no-load voltage, torque, inductance, unbalanced magnetic force and losses—will be derived and used in the design optimization. It is important to stress that the purpose of this chapter is not to offer major contribution in machine modeling, but rather to distinguish and model most important EM parameters of high-speed PM machines. The modeling does not look for ultimate accuracy but for *adequate* representation of dominant phenomena that can serve as a good basis for designing high-speed PM machines and assessing their limits.

The chapter starts with presenting model geometry and properties that will be used for modeling of slotless PM machine. Section 3.3 presents the analytical modeling of the magnetic field that forms the basis for most of models in the rest of the chapter. Main EM parameters of the machine—no-load voltage, torque, inductance—are presented in Sect. 3.4. Models for unbalanced magnetic force in a PM machine are investigated in Sect. 3.5 and comprehensive modeling of different frequency-dependent losses is given in Sect. 3.6.

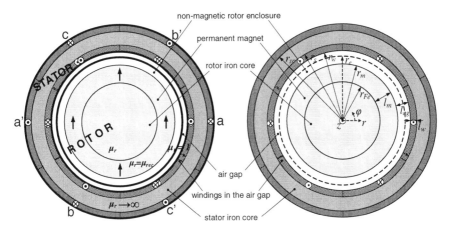

Fig. 3.1 Cross-section of the slotless PM machine (twice)

3.2 Model Geometry and Properties

A slotless machine is adequately represented as a compound of several coaxial cylinders which renders the machine suitable for modeling in a cylindrical system. The machine will be modeled in 2D with all the variables dependent only on radial and circular coordinate (r and φ); all end-effects will be neglected. The cross-section of the modeled machine is represented twice in Fig. 3.1. The machine configuration is depicted on the left-hand side and variables which define the model geometry are shown on the right.

Material properties of all machine parts are assumed to be stationary, isotropic and linear; the last condition facilitates use of superposition of the field vectors resulting from different sources. Both iron parts in the model have infinite permeability. The magnet working point is assumed to be in the second quadrant of the BH plane and recoil permeability of the magnet is μ_{rec}. All other machine parts have relative permeability equal to 1.

The most inner rotor part is the iron shaft on which permanent magnet ring resides. The magnet has two poles as it is usual with very high-speed machines. The PM ring is magnetized diametrically[1] yielding the following expression for remanent flux density:

$$\overrightarrow{B_{rem}} = \hat{B}_{rem} \sin{(\varphi_r)} \overrightarrow{i_r} + \hat{B}_{rem} \cos{(\varphi_r)} \overrightarrow{i_\varphi}, \qquad (3.1)$$

where $\hat{B}_{rem}$ is the arithmetic value of the magnet remanent flux density.

In reality, the magnet is retained with a non-magnetic enclosure/sleeve; however, since the sleeve has, practically, the same magnetostatic properties as air, that region will not be separately treated in the modeling. Thus, external rotor radius r_e and the

---

[1] The magnetization is equivalent to the ideal Halbach array with two poles.

Fig. 3.2 Angular density of conductors of the phase a

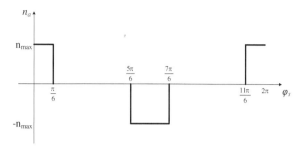

sleeve thickness are irrelevant for the EM modeling and in the further text in this chapter the term *air gap* will actually determine mechanical air gap and rotor sleeve combined.

The stator has a slotless iron core. The conductors are shown in Fig. 3.1 to be toroidally wound around the stator core, as it is the case of the test machine. However, the magnetic field within the machine is the same for the case of, more common, air-gap windings since only parts of the windings that are inside of the air gap matter for the magnetic field in this model. Furthermore, magnetic field outside of the stator core is neglected and the stator surface at $r = r_{so}$ is the surface with a flux-parallel boundary condition.

Distribution of the conductors of the phase a of the test machine is given in Fig. 3.2 in terms of angular conductor density (number of conductors per radian) over the stator circumference. The distribution can be decomposed in a Fourier series as follows:

$$n_a(\varphi_s) = \sum_{k=1,3,5\ldots}^{\infty} n_k \cos(k\varphi_s) = \frac{4}{\pi} n_{max} \sum_{k=1,3,5\ldots}^{\infty} \frac{\sin k\frac{\pi}{6}}{k} \cos(k\varphi_s), \qquad (3.2)$$

where:

$$n_{max} = \frac{3N}{\pi}; \qquad (3.3)$$

N is the number of turns per phase.[2]

The conductor distributions of the phases b and c are phase shifted for $\pm 120°$:

$$n_b(\varphi_s) = n_a\left(\varphi_s - \frac{2\pi}{3}\right) \qquad (3.4)$$

$$n_c(\varphi_s) = n_a\left(\varphi_s + \frac{2\pi}{3}\right) \qquad (3.5)$$

---

[2] In the example of toroidal windings N is physically a *half* number of turns per phase.

Fig. 3.3 A position vector
in the stationary and rotating
reference frame

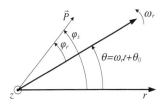

Alternatively, the conductor distribution (3.2) can be represented in an equivalent form:

$$n_a(\varphi_s) = \sum_{k=-\infty}^{\infty} n_{6k+1} \cos(k\varphi_s) + \sum_{k=0}^{\infty} n_{6k+3} \cos(k\varphi_s) \tag{3.6}$$

Finally, it will be assumed that the rotor revolves in synchronism with the stator field, thus:

$$\omega_r = \omega_e = \omega, \tag{3.7}$$

where ω_r and ω_e are mechanical (rotor) and electrical angular frequency, respectively.

The field quantities in the model will be represented in one of two reference frames: the stationary frame, fixed to the stator, and the rotating frame, fixed to the rotor. Correlation between these frames is depicted in Fig. 3.3. Angular positions in two coordinate systems are correlated in time as:

$$\varphi_s = \varphi_r + \theta, \tag{3.8}$$

where $\theta = \theta(t)$ determines the rotor angular position in stationary reference frame:

$$\theta(t) = \omega t + \theta_0. \tag{3.9}$$

In (3.9) θ_0 is the rotor initial position.

3.3 Modeling of the Magnetic Field

The field in the machine will be solved using magnetic vector potential $\overrightarrow{A}$. The vector potential is implicitly defined by its curl and divergence:

$$\overrightarrow{B} = \nabla \times \overrightarrow{A}, \tag{3.10}$$

$$\nabla \cdot \overrightarrow{A} = 0; \tag{3.11}$$

the latter expression is known as the *Coulomb gauge*.

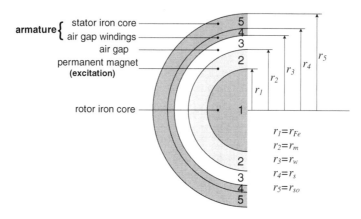

Fig. 3.4 Model geometry

As mentioned in the beginning of the chapter, the modeling will conform to the magnetostatic assumption and the effect of induced eddy currents on the field will be neglected. Therefore, Ampere's law simply yields:

$$\nabla \times \vec{H} = \vec{J_s},$$ (3.12)

where J_s is current density of the stator (air-gap) conductors.

Combining (3.12) and constitutive equation for magnetic field in a linear medium:

$$\vec{B} = \mu \vec{H} + \overrightarrow{B_{rem}}$$ (3.13)

and by incorporating (3.10) and (3.11), one can obtain the governing equation for the magnetic vector potential of a magnetostatic field in the slotless PM machine:

$$\Delta \vec{A} = -\mu \vec{J_s} - \nabla \times \overrightarrow{B_{rem}}.$$ (3.14)

The symbol $\Delta = \nabla^2$ is the Laplace operator and $\mu = \mu_0 \mu_r$.

The Eq. (3.14) is a Poisson's differential equation and it will be used for obtaining expressions of the magnetic field quantities. The equation will be solved using the principle of superposition: two terms on the right-hand side of (3.14) can be considered as *sources* of the field and the resulting vector potentials from each source can be separately determined and then summed. Finally, after solving (3.14) for the vector potential, magnetic flux density $\vec{B}$ and field intensity $\vec{H}$ can be simply obtained from (3.10) and (3.13).

The governing equation (3.14) is the general one and it is valid for the whole machine from Fig. 3.1. However, the equation takes on different, simpler forms in different machine regions. The model geometry defined in the previous section can be subdivided into 5 regions with different magnetic properties and/or governing

Table 3.1 Properties and field equations of the modeled regions

i	Region	Range for r	μ_r	Governing equation	Constitutive equation
1	Rotor iron	$0 \le r < r_{Fe}$	∞	$\Delta \vec{A} = 0$	$\vec{H} = 0$
2	Magnet	$r_{Fe} \le r < r_m$	μ_{rec}	$\Delta \vec{A} = -\nabla \times \overrightarrow{B_{rem}}$	$\vec{H} = (\vec{B} - \overrightarrow{B_{rem}})/\mu_0\mu_{rec}$
3	Air gap	$r_m \le r < r_w$	1	$\Delta \vec{A} = 0$	$\vec{H} = \vec{B}/\mu_0$
4	Windings	$r_w \le r < r_s$	1	$\Delta \vec{A} = -\mu_0 \vec{J}_s$	$\vec{H} = \vec{B}/\mu_0$
5	Stator iron	$r_s \le r < r_{so}$	∞	$\Delta \vec{A} = 0$	$\vec{H} = 0$

equations. The regions are enumerated for easier representation in the equations and represented in Fig. 3.4. The governing equations of different regions are listed in Table 3.1.

Finally, the governing equation can be also written in a scalar form. It is evident (Fig. 3.1) that current density vector has only a z-component and the same is true for a curl of the remanence vector. Therefore, magnetic vector potential has also only z-component:

$$\vec{A} = A_z (r, \varphi) \cdot \vec{i_z} \qquad (3.15)$$

and the Eq. (3.14) is equivalent to:

$$\frac{\partial^2 A_z}{\partial r^2} + \frac{1}{r^2} \frac{\partial^2 A_z}{\partial \varphi^2} + \frac{1}{r} \frac{\partial A_z}{\partial r} = -\mu J_{s,z} - \frac{B_{rem,\varphi}}{r} - \frac{\partial B_{rem,\varphi}}{\partial r} + \frac{1}{r} \frac{\partial B_{rem,r}}{\partial \varphi}. \quad (3.16)$$

The radial and angular components $B_{rem,r}$ and $B_{rem,\varphi}$ of the remanent flux density vector $\overrightarrow{B_{rem}}$ were defined in (3.1).

Boundary conditions for solving equation proceed from Ampere's and flux-conservation law and may be defined as follows:

$$\vec{n} \cdot \left(\vec{B_i} - \overrightarrow{B_{i+1}} \right) = 0, \qquad (3.17)$$

$$\vec{n} \times \left(\vec{H_i} - \overrightarrow{H_{i+1}} \right) = \vec{K_i}, \qquad (3.18)$$

where $\vec{n}$ is a unit vector normal on the boundary surface and directed from the region i to the region $i + 1$ while $\vec{K_i}$ is the surface current density at the boundary surface.

For the example of the modeled machine these boundary conditions could be rewritten in simpler forms:

$$B_{i,r} (r = r_i) = B_{i+1,r} (r = r_i), \quad i = 1..5, \qquad (3.19)$$

$$H_{i,\varphi} (r = r_i) = H_{i+1,\varphi} (r = r_i), \quad i = 1..5. \qquad (3.20)$$

There is no magnetic field outside the modeled machine regions given in Fig. 3.4, thus $\overrightarrow{B_6} = 0$ and $\overrightarrow{H_6} = 0$.

3.3.1 Field of the Permanent Magnet

The field from the permanent magnet, or the excitation field, will be modeled first. The governing equation for this case reduces to:

$$\Delta \overrightarrow{A} = -\nabla \times \overrightarrow{B_{rem}}. \tag{3.21}$$

It can be easily shown that the curl of the remanent field vector defined as (3.1) is zero:

$$\nabla \times \overrightarrow{B_{rem}} = 0, \tag{3.22}$$

and that further reduces the governing equation (3.21) to the Laplace's equation:

$$\Delta \overrightarrow{A} = 0. \tag{3.23}$$

With respect to the type of the magnetization, the solution for the magnetic potential can be found in the following form:

$$\begin{aligned}
\overrightarrow{A_1} &= -C_1 r \cos(\varphi_r) \cdot \overrightarrow{i_z}, \\
\overrightarrow{A_i} &= -\left(C_i r + \frac{D_i}{r}\right) \cos(\varphi_r) \cdot \overrightarrow{i_z}, \quad i = 2..5
\end{aligned} \tag{3.24}$$

Magnetic flux density can then be calculated from (3.10) as:

$$\begin{aligned}
\overrightarrow{B_1} &= C_1 \sin(\varphi_r) \cdot \overrightarrow{i_r} + C_1 \cos(\varphi_r) \cdot \overrightarrow{i_\varphi}, \\
\overrightarrow{B_i} &= \left(C_i + \frac{D_i}{r^2}\right) \sin(\varphi_r) \cdot \overrightarrow{i_r} + \left(C_i - \frac{D_i}{r^2}\right) \cos(\varphi_r) \cdot \overrightarrow{i_\varphi}, \quad i = 2..5
\end{aligned} \tag{3.25}$$

and magnetic field intensity $\overrightarrow{H}$ is calculated in each region from the corresponding constitutive equation.

Boundary coefficients C_i, D_i were calculated symbolically in Matlab from the system of boundary conditions (3.19) and (3.20). The final analytical expressions for the field vectors are rather lengthy and will not be presented here.

Results of the modeling of the PM field are presented for the test motor (the parameters are shown in Table 3.2). Distribution of radial and tangential flux density at various radii in the machine are given in Fig. 3.5 and 3.6 and flux lines are shown in Fig. 3.7. As expected, diametrical magnetization in the rotor causes a perfectly sinusoidal spatial distribution of the magnetic field throughout the whole machine. Analytical model calculations match results of 2D FEM modeling.

Table 3.2 Parameters of the test machine

Parameter	Symbol	Value
Rotor shaft radius	r_{fe}	10.5 mm
Magnet outer radius	r_m	14.5 mm
Winding region inner radius	r_w	17.5 mm
Stator inner radius	r_s	18.5 mm
Stator outer radius	r_{so}	27.3 mm
Magnet remanent flux density	$\hat{B}_{rem}$	0.504 T
Magnet recoil permeability	μ_{rec}	1.15

Fig. 3.5 PM field: spatial distribution of the radial flux density; the *dashed lines* represent results of the corresponding 2D FEM model

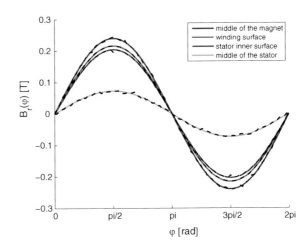

3.3.2 Armature Field

The field of the stator currents—armature field—will be modeled in this subsection. The governing equation for this case yields:

$$\Delta \vec{A} = -\mu \vec{J_s},$$

$$(3.26)$$

where $\vec{J_s}$ denotes the vector of total current density of the stator conductors which reside in the air-gap winding region. This vector has only the z-component:

$$\vec{J_s} = J_s \cdot \vec{i_z}.$$

$$(3.27)$$

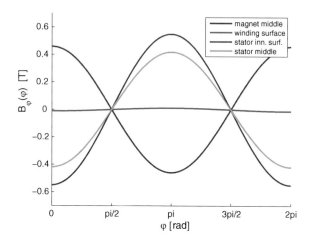

Fig. 3.6 PM field: spatial distribution of the tangential flux density

Fig. 3.7 Flux lines of the field of the permanent magnet (contour plot)

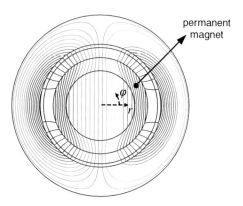

The stator current density J_s is calculated by summing the current density of each phase in the following way:

$$J_s = J_s\,(\varphi_s, t) = \frac{[n_a\,(\varphi_s)\,i_a\,(t) + n_b\,(\varphi_s)\,i_b\,(t) + n_c\,(\varphi_s)\,i_c\,(t)]}{r_{wc}l_w}, \qquad (3.28)$$

where i_a, i_b and i_c are phase currents, $l_w = r_s - r_w$ is the winding thickness and r_{wc} is the radius at the center of the windings:

$$r_{wc} = \frac{r_w + r_s}{2}. \qquad (3.29)$$

The phase currents can be represented as sums of their time harmonics. Since currents of a balanced three-phase system do not contain even and triplen harmonics [6], the phase currents can be represented as:

$$i_a = \sum_{m=-\infty}^{\infty} \hat{i}_{6m+1} \cos\left[(6m+1)\,\omega t\right]$$

$$i_b = \sum_{m=-\infty}^{\infty} \hat{i}_{6m+1} \cos\left[(6m+1)\left(\omega t - \frac{2\pi}{3}\right)\right] \qquad (3.30)$$

$$i_c = \sum_{m=-\infty}^{\infty} \hat{i}_{6m+1} \cos\left[(6m+1)\left(\omega t + \frac{2\pi}{3}\right)\right]$$

Using trigonometric identities and Eqs. (3.28), (3.30) and (3.6) it can be shown that the current density can be represented in the following form:

$$J_s = \sum_{k=-\infty}^{\infty} \sum_{m=-\infty}^{\infty} J_{s,6k+1,6m+1}$$

$$= \frac{3}{2} \sum_{k=-\infty}^{\infty} \sum_{m=-\infty}^{\infty} \frac{n_{6k+1}\hat{i}_{6m+1}}{r_{wc}l_w} \cos\left[(6k+1)\,\varphi_s - (6m+1)\,\omega t\right] \qquad (3.31)$$

The governing equation takes on the full form (3.26) only in the winding region:

$$\Delta\overrightarrow{A_4} = -\mu J_s, \qquad (3.32)$$

while it reduces to the Laplace's equation in other regions:

$$\Delta\overrightarrow{A_i} = 0, \quad i = 1, 2, 3, 5 \qquad (3.33)$$

Magnetic vector potential is found in the following form:

$$\overrightarrow{A_i} = \sum_{k=-\infty}^{\infty} \sum_{m=-\infty}^{\infty} A_{i,6k+1,6m+1} \cdot \overrightarrow{i_z}, \qquad (3.34)$$

$$A_{1,6k+1,6m+1} = -C_{1,6k+1}r^{|6k+1|}J_{s,6k+1,6m+1},$$

$$A_{i,6k+1,6m+1} = -(C_{i,6k+1}r^{|6k+1|} + D_{i,6k+1}r^{-|6k+1|})J_{s,6k+1,6m+1}, \quad i = 2, 3, 5$$

$$A_{4,6k+1,6m+1} = -\left(C_{4,6k+1}r^{|6k+1|} + D_{4,6k+1}r^{-|6k+1|} + \frac{\mu_0 r^2}{(6k+1)^2 - 4}\right)J_{s,6k+1,6m+1}.$$

$$(3.35)$$

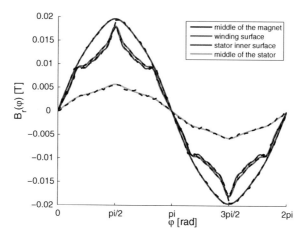

Fig. 3.8 Armature field: spatial distribution of the radial flux density; the *dashed lines* represent results of the corresponding 2D FEM model

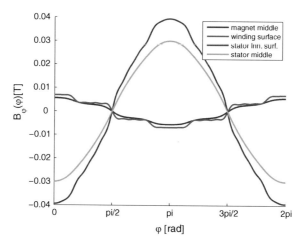

Fig. 3.9 Armature field: spatial distribution of the tangential flux density

Magnetic flux density is then obtained using (3.10) and boundary coefficients are again solved using the Matlab symbolic solver. Spatial distribution of radial and tangential flux density at different radii of the test machine are plotted in Figs. 3.8 and 3.9 for the maximum amplitude of the current—$2\sqrt{2}$ A—at the moment $t = 0$; results were confirmed by 2D FEM modeling. Time harmonics of the current do not influence the field distribution and are not considered.

For the given value of the current, the magnitude of the armature field is much smaller than the magnitude of the field of the magnet even though a magnet with rather small remanence is used. The reason for this is a very large effective air gap

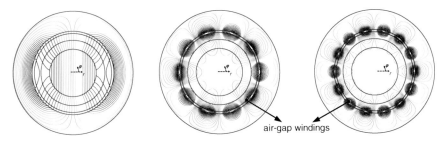

Fig. 3.10 Flux lines of the armature field: 1st , 5th and 7th spatial harmonic (contour plot)

which comprises the magnet thickness, sleeve thickness, mechanical air gap and the winding thickness. Additionally, it is apparent that higher spatial harmonics of the flux density attenuate when approaching the rotor shaft. This property is beneficial for alleviation of the rotor induced losses.

The attenuation of spatial harmonics is also demonstrated in Fig. 3.10 where field lines of the first, fifth and seventh harmonics (k equals to 0, -1 and 1, respectively) of the armature field are plotted.

3.3.3 Combined Field

Using the principle of superposition, field vectors of the combined field can be obtained by summing the field vectors resulting from the field of the permanent magnet and stator currents separately. Therefore, combined magnetic vector potential can be represented in the stator reference frame as:

$$\vec{A} = \overrightarrow{A^{pm}}(r, \varphi_s - \theta(t)) + \overrightarrow{A^{curr}}(r, \varphi_s) \tag{3.36}$$

and, similarly, magnetic flux density of the combined field becomes:

$$\vec{B} = \overrightarrow{B^{pm}}(r, \varphi_s - \theta(t)) + \overrightarrow{B^{curr}}(r, \varphi_s). \tag{3.37}$$

Magnetic vector potentials from the separate sources were derived in the previous two subsections and given by Eqs. (3.24) and (3.35). Correlation between angular coordinates of the two reference frames are given by (3.8) and (3.9).

Figure 3.11 shows distribution of the combined radial flux density for the case when stator currents field leads the magnet field for $\pi/2$. The distribution is plotted for the maximum amplitude of the currents at the moment $\omega t = \pi/2$. It is evident that the influence of the armature field on the total field in machine is practically negligible.

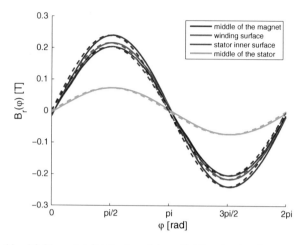

Fig. 3.11 Combined field: spatial distribution of the radial flux density, the armature field leading the permanent magnet for $\pi/2$. *Dashed lines* represent the flux density of the magnet only

3.4 Derived Quantities

3.4.1 No-Load Voltage

Flux linkage of a coil can be found as a contour integral of magnetic vector potential along the coil, thus:

$$\psi = \oint_C \vec{A} \cdot d\vec{l} . \tag{3.38}$$

Therefore, PM-flux linkage of the phase a can be calculated by:

$$\psi_a = 2l_s \int_{-\pi/2}^{\pi/2} n_a\,(\varphi_s)\,A_z^{pm}\,(r = r_{wc},\varphi_r)\,d\varphi_s, \tag{3.39}$$

or, after taking into account (3.6) and (3.8):

$$\psi_a = 2l_s \sum_{k=-\infty}^{\infty} n_{6k+1} A_z^{pm}\,(r_{wc}) \int_{-\pi/2}^{\pi/2} \cos\left[(6k+1)\,\varphi_s\right] \cos\left(\varphi_s - \theta\right) d\varphi_s. \tag{3.40}$$

In the equations above l_s represents stator axial length and $A_z^{pm}\,(r_{wc})$ denotes the part of the vector potential which is *not* dependent on φ_r (see Eq. (3.24)):

$$A_z^{pm}\,(r = r_{wc},\varphi_r) = A_z^{pm}\,(r_{wc}) \cdot \cos\left(\varphi_r\right) . \tag{3.41}$$

The integral in (3.40) has a non-zero value only when $k = 0$. Hence, the flux linkage becomes:

$$\psi_a = \pi l_s n_1 A_z^{pm} (r_{wc}) \cos \theta, \tag{3.42}$$

which, after substituting expressions for n_1 and θ (Eqs. (3.2), (3.3) and (3.9)), yields:

$$\psi_a = \frac{6}{\pi} N l_s A_z^{pm} (r_{wc}) \cos (\omega t + \theta_0) . \tag{3.43}$$

Finally, induced voltage in the phase a, no-load voltage, is obtained as the time derivative of the flux linkage:

$$e_a = \frac{d}{dt} \psi_a = -\hat{e} \sin (\omega t + \theta_0) , \tag{3.44}$$

$$\hat{e} = \left| \frac{6}{\pi} \omega N l_s A_z^{pm} (r_{wc}) \right| . \tag{3.45}$$

No-load voltage of the other two phases is, naturally, phase-shifted for $\pm 2\pi/3$ with respect to the phase a:

$$\begin{aligned} e_b &= -\hat{e} \sin \left(\omega t + \theta_0 - \frac{2\pi}{3} \right) , \\ e_c &= -\hat{e} \sin \left(\omega t + \theta_0 + \frac{2\pi}{3} \right) . \end{aligned} \tag{3.46}$$

The expression (3.45) represents the amplitude of the machine no-load voltage. In general, exact analytical expression for $\hat{e}$ is lengthy, however, it takes on a shorter form for $\mu_{rec} = 1$:

$$\hat{e}_{\mu_{rec}=1} = \frac{3}{\pi} \omega N l_s \frac{r_m^2 - r_{fe}^2}{r_s^2 - r_{fe}^2} \frac{r_s^2 + r_{wc}^2}{r_{wc}} \hat{B}_{rem} . \tag{3.47}$$

The last expression can be written in another form:

$$\hat{e}_{\mu_{rec}=1} = 2 \omega k_w N l_s r_s \frac{l_m}{g} \hat{B}_{rem} \cdot \frac{r_m + r_{fe}}{r_s + r_{fe}} \cdot \frac{1}{2} \left(\frac{r_s}{r_{wc}} + \frac{r_{wc}}{r_s} \right) , \tag{3.48}$$

where $k_w = 3/\pi$ is the machine winding factor, $l_m = r_m - r_{fe}$ is the magnet thickness and $g = r_s - r_{fe}$ is the effective air gap, or:

$$\hat{e}_{\mu_{rec}=1} = \hat{e}_{1D,\mu_{rec}=1} \cdot \frac{r_m + r_{fe}}{r_s + r_{fe}} \cdot \frac{1}{2} \left(\frac{r_s}{r_{wc}} + \frac{r_{wc}}{r_s} \right) , \tag{3.49}$$

where $\hat{e}_{1D,\mu_{rec}=1}$ is the no-load voltage would result from a one-dimensional model which does not consider curvatures of the field in the air gap.

The last term in the product on the right side of the Eq. (3.49) is very closely equal to 1, therefore, that equation can be quite accurately expressed as:

$$\hat{e}_{\mu_{rec}=1} \approx \hat{e}_{1D,\mu_{rec}=1} \frac{r_m + r_{fe}}{r_s + r_{fe}}. \tag{3.50}$$

With the last equation one can assess the accuracy of an one-dimensional model of a slotless machine. For the example of the test machine, the 1D model would overestimate the machine voltage for, approximately, 16 % with respect to the voltage which would result from more accurate 2D modeling.

3.4.2 Torque and Power

Since the developed model does not account for induced losses in the machine, the easiest way to calculate machine torque and power is from the equilibrium of input/electrical and output/mechanical power:

$$P_e = P_m, \tag{3.51}$$

or, in another form:

$$e_a i_a + e_b i_b + e_c i_c = T\omega, \tag{3.52}$$

where T is the machine torque.

Using Eqs. (3.44), (3.46) and (3.30) it is not difficult to show that expression for the machine power is:

$$P = P_e = \frac{3}{2}\hat{e} \sum_{m=-\infty}^{\infty} \hat{i}_{6m+1} \sin(6m\omega t - \theta_0). \tag{3.53}$$

It can be concluded that oscillations of power and torque of the machine is, practically, influenced solely by non-fundamental harmonics of the current. This conclusion is generally valid for slotless machines regardless of the rotor magnetization since the time harmonics of the field have predominant influence on torque oscillations and rotor losses (see [7]).

The average power is given by:

$$P_{avg} = -\frac{3}{2}\hat{e}\hat{i}\sin(\theta_0) \tag{3.54}$$

and its maximum is reached when the rotor position lags behind the maximum field of the currents for $\pi/2$:

$$P_{avg,max} = \left(\theta_0 = -\frac{\pi}{2}\right) = \frac{3}{2}\hat{e}\hat{i}. \tag{3.55}$$

This form could have also been anticipated from the sinusoidal forms of the phase currents and no-load voltages. In the equations $\hat{i} = \hat{i}_1$ is the amplitude of the phase currents.

The maximum torque is then:

$$T_{avg,max} = \frac{3}{2}\frac{\hat{e}\hat{i}}{\omega}. \tag{3.56}$$

3.4.3 Phase Inductance

Similarly to Sect. 3.4.1, linkage of the phase a with the field of the currents can be calculated by:

$$\psi_a = 2l_s \int\limits_{-\pi/2}^{\pi/2} n_a\,(\varphi_s)\,A_z^{curr}\,(r = r_{wc},\ \varphi_s)\,d\varphi_s \tag{3.57}$$

After taking into account expressions (3.6), (3.31) and (3.35), the last equation can be expressed as:

$$\psi_a = 2l_s \sum_{k_1=-\infty}^{\infty} \sum_{k_2=-\infty}^{\infty} \frac{3}{2}\frac{n_{6k_1+1}n_{6k_2+1}}{l_w r_{wc}} A_{z,6k_2+1}^{curr}\,(r_{wc}) \sum_{m=-\infty}^{\infty} \hat{i}_{6m+1}\cdots$$

$$\cdots \int\limits_{-\pi/2}^{\pi/2} \cos\left[(6k_1+1)\,\varphi_s\right]\cos\left[(6k_2+1)\,\varphi_s - (6m+1)\,\omega t\right]d\varphi_s, \tag{3.58}$$

where:

$$A_{z,6k+1}^{curr}\,(r_{wc}) = \frac{A_{z,6k+1,6m+1}^{curr}\,(r_{wc},\ \varphi_s)}{J_{s,6k+1,6m+1}\,(\varphi_s)}. \tag{3.59}$$

The integral in the last equation has a non-zero value only for $k_1 = k_2 = k$, thus the last expression can be reduced to:

$$\psi_a = 2l_s \sum_{k=-\infty}^{\infty} \frac{3}{2}\frac{n_{6k+1}^2}{l_w r_{wc}} A_{z,6k+1}^{curr}\,(r_{wc})\,\frac{\pi}{2} \sum_{m=-\infty}^{\infty} \hat{i}_{6m+1}\cos\left[(6m+1)\,\omega t\right], \tag{3.60}$$

and, with respect to (3.30), further to:

$$\psi_a = \frac{3}{2}\pi l_s \sum_{k=-\infty}^{\infty} \frac{n_{6k+1}^2}{l_w r_{wc}} A_{z,6k+1}^{curr}(r_{wc}) \cdot i_a. \tag{3.61}$$

The total (synchronous) phase inductance is than obtained as:

$$L = \frac{\psi_a}{i_a} = \frac{3}{2}\pi l_s \sum_{k=-\infty}^{\infty} \frac{n_{6k+1}^2}{l_w r_{wc}} A_{z,6k+1}^{curr}(r_{wc}). \tag{3.62}$$

The expression for inductance (3.62) comprises the self-inductance of a phase, mutual inductance and the air-gap leakage inductance since all the correlated effects have been included in the developed 2D model of the machine field. On the other hand, the expression does not account for change of the inductance with frequency since the frequency-dependent effects (eddy currents) have not been modeled.

The exact form of Eq. (3.62) is rather long and will not be presented here. The expression for inductance resulting from a 1D model, on the other hand, is simple and is given by:

$$L_{1D} = \frac{6}{\pi}\mu_0 l_s r_s \frac{(k_w N)^2}{l_m/\mu_{rec} + l_{ag} + l_w}. \tag{3.63}$$

However, this expression overestimates inductance of the test machine for about 20 %.

3.5 Unbalanced Magnetic Pull and Machine Stiffness

Asymmetric displacement of the electromagnetic components of an electrical machine brings about unbalanced magnetic pull (UMP) between the rotor and stator. Reasons for the asymmetry of PM machines can be put down to two main factors: (i) central asymmetry of the armature field due to inherently asymmetric pole/slot configuration and (ii) rotor eccentricity.[3]

The former reason is fairly irrelevant for slotless machines since their armature is rarely asymmetric; besides, the armature field is rather weak with respect to the field of permanent magnets.

On the other hand, due to irregularities in the bearings, some eccentricity of the rotor is always present. As a result, during machine's operation a revolving attraction force between rotor and stator exists which, in turn, might cause severe rotor vibrations and bearings wear. This phenomenon is particularly unwanted in machining and hard-disc-drive spindles whose applications are exceptionally susceptible to spindle runout [11, 12].

---

[3] A good part of this section has been taken from Borisavljevic et al. [10].

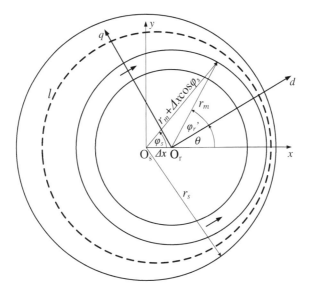

Fig. 3.12 Simplified model geometry of a slotless machine with an eccentric rotor

The attraction between stator and rotor counteracts the bearing force and appears as a machine *negative stiffness*. In the example of soft-mounted high-speed machines this effect must be considered early in the design phase of the system. Machine stiffness, thus, must be considerably lower than the stiffness of the bearings; furthermore, maximum tolerated unbalanced force must be below force capacity of the bearings.

For electromagnetic design, in most of cases, a reasonable estimation of the unbalanced pull is sufficient. Space harmonics of UMP might also be interesting for rotor-dynamical analysis [13], however, this will not be addressed in the thesis since it is of little relevance for the test machine.

Therefore, this section focuses on finding a *sufficiently* accurate model for unbalanced magnetic force in slotless PM machines. The estimation of the unbalanced force will be utilized in Chap. 9 when the requirements for the stiffness and force capacity of the bearings are set.

The geometry which will be used for modeling in this section is depicted in Fig. 3.12. The stationary reference frame xy has its origin in the center of the stator. Center of the rotor is shifted from the stator center for the displacement Δx. The rotating reference frame dq is connected to the center of the rotor with the d-axis coinciding with the rotor magnetization vector. Angular position of the rotor is defined by the angle θ which is given by Eq. (3.9). For very small displacements it is reasonable to approximate the angular position φ_r' in the rotating frame with:

$$\varphi_r' \approx \varphi_s - \theta + \frac{\Delta x}{r_m} \sin \varphi_s \approx \varphi_s - \theta. \tag{3.64}$$

For such an irregular geometry, Maxwell's stress method is arguably the easiest way to calculate the unbalanced force. The linearity assumption will be made—the effect of iron saturation can be easily neglected due to machine's very large effective air gap. In the stationary reference frame the force per unit length that acts on the rotor can be determined using the following equations [12, 14]:

$$F'_x = -\frac{1}{2\mu_0} \oint_l \left[\left(B^2_{\varepsilon,\varphi} - B^2_{\varepsilon,r} \right) \cos \varphi_s + 2 B_{\varepsilon,r} B_{\varepsilon,\varphi} \sin \varphi_s \right] r(\varphi_s) d\varphi_s, \qquad (3.65)$$

$$F'_y = -\frac{1}{2\mu_0} \oint_l \left[\left(B^2_{\varepsilon,\varphi} - B^2_{\varepsilon,r} \right) \sin \varphi_s - 2 B_{\varepsilon,r} B_{\varepsilon,\varphi} \cos \varphi_s \right] r(\varphi_s) d\varphi_s, \qquad (3.66)$$

where l is a closed surface in the air gap and $B_{\varepsilon,r}$ and $B_{\varepsilon,\varphi}$ are radial and tangential components of the magnetic flux density in the machine with eccentricity:

$$\overrightarrow{B_\varepsilon} = B_{\varepsilon,r} \cdot \overrightarrow{i_r} + B_{\varepsilon,\varphi} \cdot \overrightarrow{i_\varphi}. \qquad (3.67)$$

In order to find the unbalanced force, it is essential to adequately represent magnetic field in the machine with rotor eccentricity. One way to analytically calculate the field in the machine with eccentricity is to directly model the field in the eccentric-rotor machine. This was done, for instance, in [15], by applying perturbation method when the governing equation and boundary conditions were formulated. However, the method leads to rather complex derivation and lengthy expressions despite the truncation of the higher-order components in the solution of the governing equation.

Another, more common way, is to correlate the air-gap flux density in the machine with and without rotor eccentricity using different, indirect approaches and useful simplifications. That correlation is usually given as a function that represents normalized flux density in the eccentric-rotor machine with respect to the flux density of the perfectly concentric machine. Since it also indicates the perturbation of the air-gap permeance as a function of the rotor displacement, the function is commonly referred to as *relative permeance* function $f(r, \varphi_s)$. The resulting flux density yields:

$$\overrightarrow{B_\varepsilon} \approx f_r(r, \varphi_s) B_r \overrightarrow{i_r} + f_\varphi(r, \varphi_s) B_\varphi \overrightarrow{i_\varphi}, \qquad (3.68)$$

where B_r and B_φ denote components of the flux density in the machine without eccentricity.

To find relative permeance function conformal mapping is often used [16–18]. In [18] magnetic field which exists between two eccentric bodies representing rotor and stator of an electrical machine is determined. Each of these abstract bodies is assumed to have constant magnetic potential. Magnetic flux density of the *homopolar* field between the bodies is found after transforming the given geometry to a similar, but concentric configuration using conformal mapping. Finally, the 2D relative permeance function is found in an analytical form as a ratio between the determined flux density and the flux density in the same abstract system without eccentricity.

The relative permeance function obtained in [18] was used to perturbate flux density of an actual, concentric PM machine to obtain the flux density of the same machine with rotor eccentricity. Similar approach was adopted also in [17] and findings of these studies complied with the outcome of FE methods.

The function that modulates air-gap flux density according to the rotor displacement can be also estimated using quite simple approximations. Surprisingly, effectiveness of such an approach has rarely been explored in recent literature, with an exception of [19]. It is a straightforward guess, though, to assume that the radial flux density in the eccentric-rotor machine is inversely proportional to the effective air gap at a given radius. This assumption, in turn, leads to the following correlation:

$$\frac{B_{\varepsilon,r}}{B_r} = \frac{g_{eff}}{g_{\varepsilon,eff}}, \tag{3.69}$$

where the effective air gap of the machine without and with eccentric rotor are given by:

$$g_{eff} = \frac{l_m}{\mu_{rec}} + l_{ag}, \tag{3.70}$$

$$g_{\varepsilon,eff} = g_{eff} - \Delta x \cos \varphi_s. \tag{3.71}$$

Using similar approximation in his paper as early as in 1963, Swann [16] obtained flux density distribution in an eccentric-rotor machine which lay quite closely to the result of a rigorous method which included conformal transformation.

For small rotor displacements ($\Delta x \ll g_{eff}$) Eq. (3.69) can be further expressed in the following way:

$$B_{\varepsilon,r}(\varphi_s) = B_r(\varphi_s) \frac{1}{1 - \dfrac{\Delta x}{g_{eff}} \cos \varphi_s} \approx B_r(\varphi_s) \cdot (1 + \varepsilon \cos \varphi_s), \tag{3.72}$$

where $\varepsilon = \Delta x / g_{eff}$ represents rotor eccentricity.

Evidently, from (3.72), a very simple perturbation- or relative permeance function is obtained:

$$f = f(\varphi_s) = 1 + \varepsilon \cos \varphi_s. \tag{3.73}$$

The function f obtained using this, very simple approach is plotted in Fig. 3.13 against radial component of the 2D relative permeance function obtained using the method outlined in [18]. It is noticeable that the shape of the simple function from (3.73) is very similar to the shape of the function obtained using conformal mapping. The difference lies in the magnitude: the simple function overestimates perturbation of the flux density. Effectiveness of both permeance functions for representing the deformed flux density and force calculation will be examined in the rest of this section.

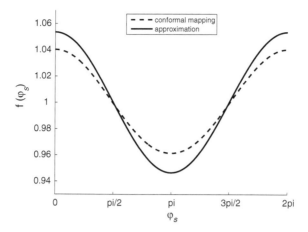

Fig. 3.13 Relative permeance function f obtained using the conformal mapping method from [18] and using approximation—Eq. (3.73)—for 10% eccentricity

The most appropriate surface for integration of the Maxwell's stress according to Eqs. (3.65) and (3.66) is the inner stator surface because at that surface the magnetic flux density in the air is strictly radial. Therefore, the Maxwell's stress at that surface reduces to:

$$\sigma_M \left(r = r_s, \varphi_s \right) = \frac{B_{\varepsilon,r}^2 \left(r = r_s, \varphi_s \right)}{2\mu_0}, \tag{3.74}$$

and, from (3.65) and (3.66), magnetic force on the rotor can be calculated by:

$$F_x' = r_s \int_0^{2\pi} \sigma_M \left(r = r_s, \ \varphi_s \right) \cos \varphi_s d\varphi_s, \tag{3.75}$$

$$F_y' = r_s \int_0^{2\pi} \sigma_M \left(r = r_s, \ \varphi_s \right) \sin \varphi_s d\varphi_s. \tag{3.76}$$

Radial flux density in the air gap at the inner stator surface of the concentric machine can be expressed as (Eq. (3.25)):

$$B_r \left(r = r_s, \varphi_s \right) = B_{4,r} \left(r = r_s, \varphi_s \right) = \left(C_4 + \frac{D_4}{r_s^2} \right) \cos \left(\varphi_s - \theta \right) \tag{3.77}$$

Radial flux density at the inner stator surface of the eccentric-rotor machine will be separately obtained using the permeance functions from Fig. 3.13. Using the simple form (3.73), the flux density in the eccentric rotor machine can be estimated as:

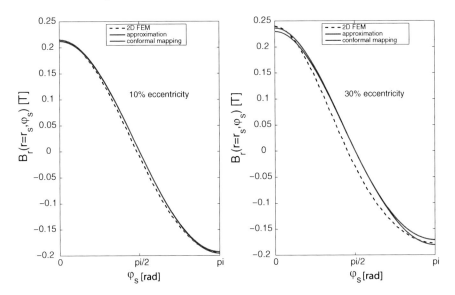

Fig. 3.14 Radial flux density at the inner stator surface of the test machine calculated by (3.78), (3.79) and 2D FEM, for 10 and 30 % eccentricity

$$B_{\varepsilon,r} = \left(C_4 + \frac{D_4}{r_s^2}\right)(1 + \varepsilon \cos \varphi_s) \cos (\varphi_s - \theta), \qquad (3.78)$$

or using the permeance function f_{cm} obtained from conformal mapping [18]:

$$B_{\varepsilon,r,cm} = \left(C_4 + \frac{D_4}{r_s^2}\right) f_{cm} \cos (\varphi_s - \theta). \qquad (3.79)$$

Adequate representation of the flux density along the stator inner surface is, thus, necessary for accurate estimation of the force. Expressions (3.78) and (3.79) are calculated for the example of the test machine when the rotor magnetization coincides with the x-axis ($\theta = 0$) and the results are compared with results from 2D FEM in Fig. 3.14.

Both permeance functions are good in representing the deformed radial flux density and the estimations are quite comparable. The distribution obtained using conformal-mapping function is slightly more accurate than the distribution when the simple correlation is used and that difference is more evident when the eccentricity is increased.

The force per unit length that acts on the rotor in the x direction can be calculated from (3.75). Using the flux density expression (3.78) and the expression for the Maxwell's stress (3.74), it can be shown, after integration, that the force is closely equal to:

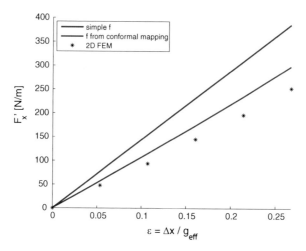

Fig. 3.15 Unbalanced force on the rotor of the test machine calculated using (i) Eq. (3.81), (ii) conformal mapping and (iii) 2D FEM

$$F'_x \approx \frac{\pi}{2\mu_0} \left(C_4 + \frac{D_4}{r_s^2} \right)^2 \frac{r_s}{g_{eff}} \left[1 + \frac{1}{2} \cos(2\theta) \right] \Delta x. \qquad (3.80)$$

As anticipated, it can be shown that the force in the y direction is equal to zero.

For this study it is important to determine the maximum unbalanced force for a given rotor displacement. From the last equation it is evident that the maximum force occurs when $\theta = k\pi$ i.e. when the rotor magnetization is parallel to the x axis and it yields:

$$F'_{x,max} = F'_x (\theta = k\pi) \approx \frac{3\pi}{4\mu_0} \left(C_4 + \frac{D_4}{r_s^2} \right)^2 \frac{r_s}{g_{eff}} \Delta x. \qquad (3.81)$$

From (3.81) machine stiffness per unit length can be expressed as:

$$k'_{PM} = - \frac{\Delta F'_{x,max}}{\Delta x} \bigg|_{\Delta x = 0} \approx - \frac{3\pi}{4\mu_0} \left(C_4 + \frac{D_4}{r_s^2} \right)^2 \frac{r_s}{g_{eff}} \qquad (3.82)$$

and it is negative because the magnetic force tends to remove the rotor from the center.

Finally, the force can be obtained in the same way just using the flux density with the relative permeance function obtained using the conformal-mapping method from [18], Eq. (3.79). This approach is fully two-dimensional since it utilizes 2D models of both relative permeance function and the magnetic field of a non-eccentric machine.

Results of both analytical models are compared with results from 2D FEM modeling for the test machine example. Both methods seem to overestimate the total force on the rotor, however, the force estimation obtained using conformal mapping is considerably more accurate. On the other hand, estimating the field using the simple permeance function is much easier. It also results in a short analytical form which offer much more insight into the unbalanced force than the complex analytical form of the function obtained from conformal mapping. From the machine design point of view $50 \div 60$ percent of overestimation resulting from the simple method is quite tolerable and can be accounted as a safety margin in the system design.

3.6 Losses in the Machine

3.6.1 Stator Core Losses

Modeling of induced losses in the stator ferromagnetic core is a rather difficult task for a number of reasons. Losses in the core stem from a combination of linear (such as eddy-currents) and non-linear phenomena (such as hysteresis) whose combined effect cannot be accurately predicted by superposition. Besides, hysteresis, as an inherently non-linear phenomenon, does not easily lend itself to an analytical modeling. Finally, classical representation of core losses as a combination of hysteresis and eddy-current losses is not sufficient to account for all the losses at high frequencies. In literature, additional, rather obscured loss named *anomalous*, *stray* or *excess* loss is often introduced to account for the loss increase without much attempting to explain its physical nature [6, 20].

At the same time, it is extremely difficult to verify models for a particular loss mechanism since it is practically impossible to separate different loss phenomena. In order to get figures on losses that can be utilized in modeling, designers usually rely on empirical correlations rather than on classical EM modeling. Empirical formula that is regularly used to calculate the core losses in an iron core is the Steinmetz equation which will be represented here in the following form:

$$p_{Fe} = C \hat{B}_m^\alpha f^\beta, \tag{3.83}$$

where p is a mass (or volume) density of the total loss power, $\hat{B}_m$ and f are flux density amplitude and frequency of the magnetic field in the core, respectively, and C, α and β are empirically obtained parameters.

However, when the core is subjected to a field whose frequency varies throughout a broad range of values, it is difficult to compose a single equation which can represent the losses accurately enough for all possible frequencies and flux densities. Furthermore, the whole loss modeling depends on (availability of) the manufacturer's data which are often incomplete and/or inadequate for the desired operating mode.

In this section both classical and empirical approaches will be taken in calculating the core losses and shown for the test machine case. The purpose of the section is to find an adequate analytical expression for power of the losses that can be used for the machine optimization.

Hysteresis loss represents the energy that is dissipated in the material while flux density $\vec{B}$ and field intensity $\vec{H}$ oscillate in cycles over the hysteresis loop. Volume density of the energy dissipated in each cycle is proportional to the area encircled by the hysteresis loop in the BH characteristic of the material. Therefore, the power density of the loss is equal to:

$$p_{Fe,h} = A_{BH} f, \tag{3.84}$$

where A_{BH} is the area of the hysteresis loop in J/m^3.

Dependence of the hysteresis-loop area on the flux density is rather complex and is property of the material; it may be expressed, however, using the approximate correlation:

$$A_{BH} \approx C_h B_m^{\alpha_h}, \tag{3.85}$$

where the parameters C_h and α_h can be adjusted according to magnetic properties of the material.

Therefore, the equation for power density of hysteresis loss becomes:

$$p_{Fe,h} \approx C_h \hat{B}_m^{\alpha_h} f. \tag{3.86}$$

For the silicon-steel core used in the test machine the parameters of the hysteresis loss were obtained from a plot of hysteresis loops given by the manufacturer—Appendix C. The hysteresis loss density is roughly estimated as:

$$p_{Fe,h} \approx 70 \hat{B}_m^{1.8} f \ [\text{W/m}^3] \tag{3.87}$$

where B_m and f are expressed in SI units.

Eddy current loss in a laminated core can be estimated from a simplified model of eddy currents in thin laminations induced by a one-dimensional field. The equation for the eddy current loss density is given by [7, 20]:

$$p_{Fe,e} = \frac{\pi^2 f^2 d^2 \hat{B}_m^2}{6 \rho_{Fe}}, \tag{3.88}$$

where d is the lamination thickness and ρ_{Fe} is resistivity of the core material.

Total power of losses can be calculated by integrating (3.86) and (3.88) over the stator core volume. For the modeled machine (Fig. 3.1), however, it is appropriate to express the flux density amplitude B_m as an equivalent flux density in the stator yoke B_y [7]:

$$\hat{B}_m = \hat{B}_y = \frac{\Phi_{max}}{(r_{so} - r_s) l_s}, \tag{3.89}$$

where Φ_{max} is the maximum flux in the stator yoke and l_s is the stator axial length.

Maximum flux can be calculated as:

$$\Phi_{max} = -l_s \int_{r_s}^{r_{so}} B_{5,\varphi}(r, \varphi = 0)\, dr \qquad (3.90)$$

and, using (3.25), it can be shown that:

$$\hat{B}_y = B_{5,\varphi}\left(r = \sqrt{r_s r_{so}}\right) = -C_5 + \frac{D_5}{r_s r_{so}}. \qquad (3.91)$$

After substituting (3.89) into (3.86) and (3.88) the total loss in the core calculated in the classical way can be simply estimated as:

$$P_{Fe} = P_{Fe,h} + P_{Fe,e} = (p_{Fe,h} + p_{Fe,e})V_s, \qquad (3.92)$$

where V_s is the stator core volume:

$$V_s = (r_{so}^2 - r_s^2)\pi l_s. \qquad (3.93)$$

Another approach to model the losses is to adjust the parameters of the Steinmetz equation (3.83) so that its results fit the manufacturer's data. Loss curves for the used core material of the test machine are obtained from the manufacturer—Appendix C (Fig. 3.15).

The parameters of the equation were first fitted using all the available data from the material loss curves for the frequency range $[50 \div 10\,\mathrm{k}]$ Hz and flux density range $[0.1 \div 1]$ T:

$$P_{Fe} \approx 7.0 \cdot 10^{-4} \hat{B}^{1.75} f^{1.50} \cdot V_s \,[\mathrm{W}]. \qquad (3.94)$$

Since, for a high-speed machine, it is most important to adequately represent losses in the high-frequency operating region, another estimate was made using loss curves for 2 and 5 kHz only (see Fig. C.2). Around those frequencies the loss approximation yields:

$$P_{Fe,hf} \approx 4.7 \cdot 10^{-4} \hat{B}^{1.86} f^{1.53} \cdot V_s \,[\mathrm{W}] \qquad (3.95)$$

Results of the models presented in this section are plotted in Fig. 3.16 for the stator core of the test machine. The figure shows estimation of the loss power versus frequency for the actual machine whose equivalent flux density in yoke was calculated by (3.91).

The discrepancy between loss approximation based on classical modeling and manufacturer's data is huge; classical modeling evidently fails to model core losses for the high-frequency operation of the machine. Not only that the non-modeled (*excess*) losses become significant at high frequencies, but also manufacturing processes which are used to assemble machine cores, such as cutting and punching, have a great influence on the final core properties [21].

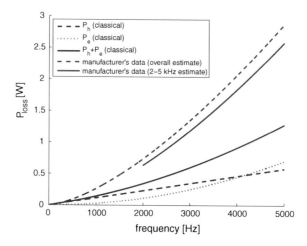

Fig. 3.16 Power of the core losses in the stator of the test machine, $\hat{B}_y = 0.43\ T$, resulting from the different models

Additionally, the plot shows considerable discrepancy between results of the Steinmetz equation fitted to all the available loss curves and the equation fitted to the specific frequency region. The Steinmetz equation, in its original form, can hardly account for iron losses at a great range of frequencies and magnetic fields. Therefore, for the machine design it is the most effective if the losses are represented only in the frequency range of the most interest. For the example of the test machine Eq. (3.95) will be used for the machine optimization in Chap. 7.

3.6.2 Copper Losses

Pulsating magnetic field of both rotor magnets and conductors current of a high-speed machine will influence the distribution of the current within the conductors and will cause formation of eddy currents in the copper. The higher the frequency of the magnetic field is the stronger its influence on losses in the conductors will be. As a result, at very high electric frequencies the total losses in conductors can significantly differ from the standard I^2R conduction loss. In this section, losses in the copper of a slotless PM machine will be assessed with particular focus on their frequency-dependent part.[4]

Skin- and proximity effect are two main sources of frequency-dependent copper losses in power transformers and they were studied in detail in literature. In works of

---

[4] A good part of this Section has been published in Borisavljevic et al. [22].

Ferreira [23–25] it was shown that skin- and proximity effect can be independently treated, due to their orthogonality. Additionally, useful expressions for the correlated losses were derived and verified and those will be the starting point for the study in this section.

Copper losses of a slotless PM machine will be divided here into three parts: (i) the conduction loss part $P_{Cu,skin}$, which also includes a rise in loss caused by the reduction of effective conductors cross-section due to skin-effect; (ii) the proximity loss part $P_{Cu,prox}$, which accounts for eddy-current loss in the conductors due to the pulsating magnetic field of the neighboring conductors; and (iii) $P_{Cu,eddy}$, which accounts for the eddy-current loss due to the pulsating magnetic field of the rotor magnet:

$$P_{Cu} = P_{Cu,skin} + P_{Cu,prox} + P_{Cu,eddy}. \tag{3.96}$$

The total conduction loss can be expressed in the following way [23, 25]:

$$P_{Cu,skin} = F(\phi) \cdot P_{Cu,DC} = \underbrace{(F(\phi) - 1) I^2 R_{DC}}_{\text{skin-effect}} + \underbrace{I^2 R_{DC}}_{\text{DC}}. \tag{3.97}$$

In Eq. (3.97) the increase of conduction losses as the result of skin effect is distinguished from the regular DC conduction loss. Function $F(\phi)$ was derived in [23, 25] as:

$$F(\phi) = \frac{\phi}{2} \cdot \frac{(ber(\phi)bei'(\phi) - bei(\phi)ber'(\phi))}{ber'^2(\phi) + bei'^2(\phi)}, \tag{3.98}$$

where parameter ϕ is proportional to the ratio between conductor diameter and the skin-depth:

$$\phi = \frac{d_{Cu}}{\delta_{skin}\sqrt{2}} = d_{Cu}\sqrt{\frac{\pi \sigma_{Cu} \mu_0 f}{2}}. \tag{3.99}$$

The second and the third term in Eq. (3.96) take on the following forms for the example of a slotless machine [23, 25]:

$$P_{Cu,prox} = \int_{l_{Cu}} \frac{G(\phi)}{\sigma_{Cu}} \cdot \hat{H}_c^2 dl, \tag{3.100}$$

$$P_{Cu,eddy} = \int_{l_{Cu,ag}} \frac{G(\phi)}{\sigma_{Cu}} \cdot \hat{H}_m^2 dl, \tag{3.101}$$

where $\hat{H}_c$ and $\hat{H}_m$ are amplitudes of magnetic field intensity in the conductor due to neighboring conductors and rotor permanent magnet, respectively; l_{Cu} is the total conductor length and $l_{Cu,ag}$ is the total conductor length in the air gap.

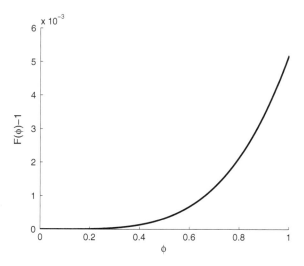

Fig. 3.17 Plot of the function $(F(\phi) - 1)$

Function $G(\phi)$ is also derived in [23, 25] as:

$$G(\phi) = 2\pi\phi \cdot \frac{(ber_2(\phi)ber'(\phi) - bei_2(\phi)bei'(\phi))}{ber^2(\phi) + bei^2(\phi)}. \tag{3.102}$$

In a slotless PM machine the field of the conductors is very small compared to the field of the permanent magnet, therefore, proximity-effect loss can be neglected. Eddy-current loss in conductors, when end effects are neglected, can be simply calculated as:

$$P_{Cu,eddy} = \frac{G(\phi)}{\sigma_{Cu}} \cdot \hat{H}_m^2 l_{Cu,ag}. \tag{3.103}$$

In order to assess the influence of frequency-dependent losses on the total copper loss in the machine, it is necessary to analyze the actual values of the functions $F(\phi)$ and $G(\phi)$.

It is not reasonable to expect conductor diameters of high-speed machines larger than 1 mm and (fundamental) electrical frequencies higher than 10 kHz. For those two particular values, at room temperature, parameter ϕ can be calculated from (3.99) to be 1.05. Therefore, it is sufficient to analyze the functions F and G for parameter ϕ in the range $[0, 1]$.

Function $(F(\phi) - 1)$ is plotted in Fig. 3.17 for the given range of ϕ. It is evident from the values of the plotted function that the skin-effect part of Eq. (3.97) certainly comprises less than one percent of the total value of $P_{Cu,skin}$ and can thus be neglected. Hence:

$$P_{Cu,skin} \approx P_{Cu,DC} = I^2 R_{DC}. \tag{3.104}$$

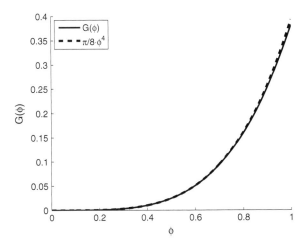

Fig. 3.18 Plot of the function $G(\phi)$ and of its approximation

In Fig. 3.18 function $G(\phi)$ is presented. Although it has a complex analytical expression (Eq. (3.102)), the function can be very well approximated with the following function:

$$G(\phi) \approx \frac{\pi}{8} \phi^4 = \frac{\pi^3 \sigma_{Cu}^2 \mu_0^2 f^2 d_{Cu}^4}{4}, \tag{3.105}$$

which was also shown in Fig. 3.18.

After inserting expression (3.105) into (3.103) and knowing that $\hat{B}_m = \mu_0 \hat{H}_m$, one can obtain the approximate expression for the eddy-current loss in the air-gap conductors:

$$P_{Cu,eddy} = \frac{\hat{B}_m^2 (2\pi f)^2 d_{Cu}^2 \sigma_{Cu}}{32} V_{Cu,ag}, \tag{3.106}$$

which is sometimes used in literature on machines (e.g. [26]). In (3.106) $V_{Cu,ag}$ is the total volume of conductors in the air gap:

$$V_{Cu,ag} = n \frac{d_{Cu}^2}{4} \pi l_{Cu,ag}, \tag{3.107}$$

where n is a number of parallel conductors in a phase.

According to (3.25) amplitude of flux density in the air-gap windings can be calculated as:

$$\hat{B}_m \approx B_{4,r}(r = r_{wc}) = C_4 + \frac{D_4}{r_{wc}^2}. \tag{3.108}$$

Finally, the total copper loss in a slotless PM machine can be calculated as:

$$P_{Cu} \approx I^2 R_{DC} + P_{Cu,eddy}, \tag{3.109}$$

where the total DC resistance is given by:

$$R_{DC} = \frac{4\rho_{Cu}}{nd_{Cu}^2\pi}l_{Cu}.$$ (3.110)

Although DC resistance does not depend on frequency it is strongly dependent on temperature since copper resistivity changes significantly with temperature.

For a toroidally wound machine it holds:

$$l_{Cu} = [2l_s + (r_{so} - r_s)\pi]6N,$$ (3.111)

$$l_{Cu,ag} = l_s6N,$$ (3.112)

where $2N$ is the number of turns per phase.

The optimization of stator conductors is presented in Sect. 7.4.3 of Chap. 7.

3.6.3 Air-Friction Loss

Both prediction and measurement of shear stress, loss and temperature increase in rotor as result of air friction on the rotor surface are important, but very difficult and thankless tasks. The importance of estimation of air-friction loss in a high-speed machine lies in the strong dependence of the loss on rotational frequency. Although virtually immaterial at low speeds, air-friction drag has the highest rate of increase with rotational speed in comparison with other loss factors and it will inevitably take a predominant portion of overall losses as the machine speed rises.

Therefore, in design of very-high-speed machines good assessment of air-friction loss is essential. Not only does air friction greatly influence the drag torque at high speeds, it also affects temperature in the rotor. Excessive rise of the rotor temperature is unacceptable because the rotor is, usually, barely cooled.

Flow of air (fluid) in the machine gap and interaction between the rotor and the air is very complex for modeling and those phenomena remain elusive for majority of machine designers. Although laminar air flow can be analytically modeled, that type of flow is dominant only at very low speeds at which air friction is fairly irrelevant. Air friction becomes important when turbulences and vortices act in the air gap and modeling of those requires, at least, comprehensive knowledge of fluid dynamics. For practical needs of engineers, however, empirical correlations are satisfactory and they are greatly utilized for estimating air-friction loss and thermal convection in the air gap. These correlations rely on empirically-obtained expressions for coefficients such as friction coefficients or convection coefficient.

It is the empirical coefficients, however, and their usage in academic papers that makes modeling of air friction rather dubious. Either there is lack of recent studies on empirical coefficients for friction-correlated parameters or today's researchers of electrical machines are not aware of them. Coefficients, which are obtained several decades ago and whose forms are too unfamiliar to have any analytical value, are repeatedly used in literature on machines, copied from one paper to another, disregarding the experimental context in which the coefficients are actually obtained. When authors, after introducing several empirical coefficients, without explanation and often missing to refer to their original source, obtain results which perfectly match analytical predictions, the results of such studies are, at most, suspicious.

Based on a few studies that give some reasonable practical evaluation of the used empirical coefficients for high-speed rotors, this section will analyze power of air-friction loss in the test machine. Since the author was not able to seriously examine validity of different models, this section will only offer a rough prediction of the total air-friction loss rather than rigorous modeling.

The power that is required from machine to overcome air-friction drag at the cylindrical surface of the rotor is given by (e.g. [27]):

$$P_{af} = k_r C_f \rho \pi \omega^3 r^4 l, \tag{3.113}$$

where ρ is the air mass density, $\omega = 2\pi f$ is rotor angular frequency, r and l are external radius and axial length of the cylinder, respectively. The coefficient C_f is called *friction coefficient* and is obtained empirically.[5] Parameter k_f represents roughness coefficient and is equal to 1.0 for perfectly smooth surfaces.

For air-friction loss at faces of a rotating disc the following equation is used:

$$P_{af} = \frac{1}{2} k_r C_f \rho \omega^3 (r^5 - r_a^5), \tag{3.114}$$

where r_a is the inner radius of the disc.

Expressions for friction coefficients are obtained from experimental work. Although they take on analytical forms in which geometrical, kinetic and material parameters of the studied bodies and the gap fluids are correlated, those expressions offer very little, if any, insight into the air friction phenomenon. Besides, different experimental studies give different expressions for the coefficients. It is, thus, of crucial importance for trustful modeling that the used friction coefficient is applied to the same or similar conditions under which the coefficient was originally obtained.

An abundance of different friction coefficients can be found in the engineering literature; however, the amount of information on origin of those coefficients is sparse. In recent years, thesis by Saari [27] has become an indispensable, if not only, resource for air-friction modeling for machine designers. Saari gave some qualitative analysis of several experimental studies on friction coefficients including the conditions under which the coefficients were obtained.

For friction at cylindrical surfaces Saari presented, among others, friction coefficients reported by Bilgen and Boulos [28]:

---

[5] Equation (3.113) is actually derived from the definition of the friction coefficient. The coefficient is defined as the ratio between shear stress and dynamic pressure at the cylinder surface.

$$C_f = 0.515 \frac{\left(\frac{\delta}{r}\right)^{0.3}}{Re_\delta^{0.5}}, \quad 500 < Re_\delta < 10^4$$

$$C_f = 0.0325 \frac{\left(\frac{\delta}{r}\right)^{0.3}}{Re_\delta^{0.2}}, \quad 10^4 < Re_\delta \tag{3.115}$$

and Yamada [29]:

$$C_f = \frac{0.0152}{Re_\delta^{0.24}}, \quad 800 < Re_\delta < 6 \cdot 10^4. \tag{3.116}$$

In (3.115) and (3.116) Re_δ is the Reynolds number for cylinders in enclosure (stator):

$$Re_\delta = \frac{\rho \omega r \delta}{\mu}, \tag{3.117}$$

where δ is the air-gap length in the radial direction and μ is dynamic viscosity of air. The Reynolds number is generally used to determined the nature of the particular gas flow.

The aforementioned coefficients were originally obtained for liquid fluids (water and oil); however, with use of coefficients from [28], Saari's predicted correlation between the loss power at the cylindrical rotor surfaces and frequency had a very similar shape as that obtain from experiments. Relatively small discrepancy between the results was put down to inappropriate roughness coefficient.

In his articles, Aglen [30–32] used coefficients from Yamada [29] to predict the air friction loss. He conducted extensive measurements on a 70.000 rpm PM generator, including direct measurements of friction losses and calorimetric tests, and the results matched the predictions pretty well.

In their paper on a very high-speed PM motor authors of [33] used the following friction coefficients, taken from the paper of Awad and Martin [34]:

$$C_f = 0.0095 \cdot Ta^{-0.2}, \tag{3.118}$$

where Ta is the Taylor number:

$$Ta = Re_\delta \left(\frac{\delta}{r}\right). \tag{3.119}$$

Zwyssig et al. [35] used expression very similar to (3.118) for friction coefficients for turbulent flow ($Ta > 400$) and verified the air-friction loss estimation, however, up to relatively moderate speeds.

For the example of an enclosed rotating disc, expressions for the friction coefficients obtained by Daily and Nece [36] are regularly used:

Fig. 3.19 Fluid flow regimes according to Daily and Nece [36]

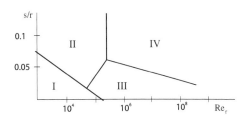

$$C_f = \frac{2\pi}{\left(\dfrac{s}{r}\right) Re_r}, \qquad \text{Regime I}$$

$$C_f = \frac{3.7 \left(\dfrac{s}{r}\right)^{0.1}}{Re_r^{0.5}}, \qquad \text{Regime II}$$

$$C_f = \frac{0.08}{\left(\dfrac{s}{r}\right)^{0.167} Re_r^{0.25}}, \qquad \text{Regime III} \qquad (3.120)$$

$$C_f = \frac{0.0102 \left(\dfrac{s}{r}\right)^{0.1}}{Re_r^{0.2}}, \qquad \text{Regime IV}$$

where s is the distance between the disc surfaces and the enclosure and Re_r is the tip Reynolds number:

$$Re_r = \frac{\rho \omega r^2}{\mu}. \qquad (3.121)$$

The different regimes of the gas flow appearing in Eqs. (3.120) are shown in Fig. 3.19. Saari also obtained good approximations for the loss using the coefficients (3.120) for end rings and discs of his test rotor.

The friction coefficients represented in this section will be applied to Eqs. (3.113) and (3.114) for example of the test machine in order to compare the different models and check their consistency. A simplified geometry of the test rotor and its setting is shown in Fig. 3.20 and the parameters are given in Table 3.3.

Since the diameter of the rotor disc is much larger than the diameter of the shaft, all models suggest that the air-friction loss at the shaft is negligible in comparison with the losses at the surfaces of the disc. Therefore, only losses at those surfaces will be presented here. The loss at the external cylindrical surface of the disc is separately calculated using Eqs. (3.115), (3.116) and (3.118), while the loss at the disc faces is calculated using (3.120). Total, combined losses resulting from these models are shown in Fig. 3.21 with respect to the rotor frequency.

From the loss plot from Fig. 3.21 it is evident that the air friction will certainly be the main source of drag and loss in the test machine. Different loss models show significant difference in prediction at very high speeds, but the rate of the loss increase with frequency is consistent. Therefore, it is necessary to take this loss into account when the power and torque of the test machine are sized.

Fig. 3.20 Simplified sketch of a half of the test-setup cross-section

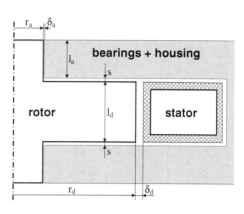

Table 3.3 Parameters of the test rotor geometry (Fig. 3.20)

Parameter	Symbol	Value
Shaft radius	r_a	4 mm
Shaft clearance	δ_a	14 µmm
Bearing length	l_a	5 mm
Rotor disc radius	r_d	16.5 mm
Machine air gap	δ_d	$\approx$1 mm
Rotor disc length	l_d	8 mm
Disc-housing clearance	s	mm

On the other hand, it is hardly possible that the air flows at the side and faces of the disc can be independently treated and their resulting effects simply summed. Furthermore, the test rotor is suspended in air bearings: compressed air from the bearings contributes significantly to the total air flow, modeling of which goes far beyond intentions of this thesis. However, the analytical expressions represented in this section can still be used for rough estimation of air-friction loss that is so important for the machine design.

3.6.4 Rotor Loss

In Sect. 3.3.2 magnetic vector potential of the armature field in a slotless PM machine was expressed as:

$$\overrightarrow{A_i} = \sum_{k=-\infty}^{\infty} \sum_{m=-\infty}^{\infty} A_{i,6k+1,6m+1}\,(r,\varphi_s) \cdot \overrightarrow{i_z}\,, \tag{3.122}$$

where algebraic values of the vector-potential harmonics were defined by Eqs. (3.31) and (3.35).

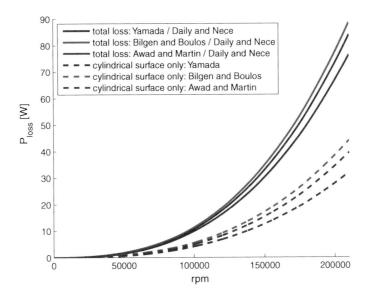

Fig. 3.21 Air-friction loss of the test machine vs. rotational speed; predictions formed using friction coefficients (3.115), (3.116), (3.118) and (3.120)

Following the derivation from Sect. 3.3.2, harmonic components of the vector potential can also be formulated in the following way:

$$A_{i,6k+1,6m+1}(r, \varphi_s)$$
$$= A_{i,6k+1,6m+1}(r) \cos[(6k+1)\varphi_s - (6m+1)\omega t], \quad i = 1..5, \quad (3.123)$$

where $A_{i,6k+1,6m+1}(r)$ depends on the radial coordinate only.

If the angular coordinate φ_s in the stationary reference frame is expressed via the angular position in the rotor reference frame (Eqs. (3.8) and (3.9)):

$$\varphi_s = \varphi_r + \omega t + \theta_0 \quad (3.124)$$

and when this is applied to (3.123), harmonic components of the vector potential in the rotor reference frame are obtained:

$$A_{i,6k+1,6m+1}(r, \varphi_r)$$
$$= A_{i,6k+1,6m+1}(r) \cos[(6k+1)(\varphi_r + \theta_0) + 6(k-m)\omega t], \quad i = 1..5.$$
$$(3.125)$$

Total z component of the magnetic vector potential in the rotor reference frame is then given by:

$$A_{i,z}(r, \varphi_r) \sum_{k=-\infty}^{\infty} \sum_{m=-\infty}^{\infty} A_{i,6k+1,6m+1}(r, \varphi_r), \quad i = 1..5. \tag{3.126}$$

Vector components of corresponding magnetic flux density in the rotating frame are obtained using (3.10):

$B_{i,6k+1,6m+1,r}$

$$= -(6k+1) \frac{A_{i,6k+1,6m+1}(r)}{r} \sin\left[(6k+1)(\varphi_r + \theta_0) + 6(k-m)\omega t\right],$$

$$\tag{3.127}$$

$B_{i,6k+1,6m+1,\varphi}$

$$= -\frac{\partial A_{i,6k+1,6m+1}(r)}{\partial r} \cos\left[(6k+1)(\varphi_r + \theta_0) + 6(k-m)\omega t\right], \quad i = 1..5.$$

From the last equations it is evident that the rotor also observes traveling waves of the flux density as a result of combination of armature spatial- and time harmonics with unequal harmonic numbers: $k \neq m$. Therefore, the rotor conductive regions (in particular, permanent magnet and rotor iron, $i = 1, 2$) are prone to induction of eddy-current losses due to the existence of the armature traveling harmonics.

Based on the magnetostatic modeling of magnetic field presented in this section, it is not possible to rigorously model rotor eddy currents; hence, the corresponding loss can only be estimated using approximate expressions. In general, rotor induced loss does not comprise a great portion of the overall losses, however, it may still be a cause of rotor failure since rotors are seldom cooled.

In his thesis Polinder [6] showed different approaches for calculating rotor losses from a magnetostatic field model. For the machine defined in Sect. 3.2 it would be of particular interest to assess losses in the magnet since it is particularly exposed to the armature harmonics and rather sensitive to heat at the same time.

For a cylindrical rotor magnets the eddy-current loss can be found from the Faraday's law:

$$\nabla \times \overrightarrow{E} = -\frac{\partial \overrightarrow{B}}{\partial t}, \tag{3.128}$$

or, alternatively:

$$\rho_m \nabla \times \overrightarrow{J} = -\frac{\partial}{\partial t}\left(\nabla \times \overrightarrow{A}\right), \tag{3.129}$$

where $\overrightarrow{E} = \rho_m \overrightarrow{J}$ is electric field, ρ_m is resistivity of the magnet and $\overrightarrow{J}$ is current density vector of induced eddy currents.

In analogy with the vector potential, it can be assumed the current density vector only has the z component. Therefore, eddy-current density in the z direction can be calculated by (see [6]):

$$J_z(r, \varphi_r) = -\frac{1}{\rho_m} \frac{\partial A_z(r, \varphi_r)}{\partial t}, \qquad (3.130)$$

having, naturally, magnetic vector potential and current density expressed in the rotor coordinates.

Finally, the loss in the magnet can be calculated by integrating power density of the induced currents $p_{m,eddy} = \rho_m J_z^2$ throughout the magnet volume:

$$P_{m,eddy} = \iiint\limits_{V_m} \rho_m J_z^2(r, \varphi_r) dV_m = \frac{l_s}{\rho_m} \int\limits_{r_{fe}}^{r_m} \int\limits_{0}^{2\pi} \left(\frac{\partial A_z(r, \varphi_r)}{\partial t}\right)^2 d\varphi_r dr. \qquad (3.131)$$

However, the magnet of the test machine does not lend itself to such calculation of the magnet eddy-current loss. Namely, an injection-molded plastic-bonded magnet is used in the test rotor in which small magnet particles are blended with a plastic binder. Although magnet has a cylindrical form, its conductive component (NdFeB magnet) is powdered and spread around the whole magnet volume so it does not form a homogeneous cylinder as assumed in derivation of the last equation.

For such a magnet structure Polinder's approach for calculating loss in segmented magnet seems more appropriate [6, 8]. For such a case volume density of the magnet eddy-current loss is approximately given by:

$$p_{m,eddy,seg} \approx \frac{b_m^2}{12\rho_m}\left(\frac{dB_r}{dt}\right), \qquad (3.132)$$

where b_m is the width of the magnet segment and B_r is the radial flux density.

Unfortunately, details of the physical structure of the test-rotor magnet, in particular, size and distribution of permanent magnet particles within the magnet composite, are unknown to the author and no approximation of the magnet eddy-current loss will be made. Nevertheless, since the actual size of the magnet particles (b_m) is very small and the armature field in the magnet is rather weak it is expected that the induced loss in the magnet is negligible.

3.7 Conclusions

The chapter presents analytical electromagnetic modeling of a (high-speed) slotless PM machine. The goal of the chapter is to distinguish dominant EM phenomena in the machine and to provide adequate representation of those phenomena that can serve as a good basis for designing high-speed PM machines and assessing their limits.

Governing differential equations that represent field in machines are expressed over magnetic vector potential in a cylindrical 2D system. Geometry and properties of the model are adapted to the test machine which is a two-pole toroidally-wound slotless PM motor; the modeling can also account for other types of slotless machines. The model is *magnetostatic* because of the small influence of the reaction field of the rotor eddy-currents on the total field in the machine. Results of the analytical field models are confirmed by 2D FEM which maintains, however, the same model geometry and assumptions as the analytical model. Based on the field equations, machine parameters—no-load voltage, phase inductance, torque and power—are derived and compared to results of 1D models.

The chapter presents an original study on calculation of unbalanced magnetic pull (force) and stiffness in a PM machine with rotor eccentricity. An approximate analytical expression for distorted magnetic field in an eccentric-rotor machine is used to determine the unbalanced magnetic force and stiffness of slotless PM machines. The effectiveness of such a model in representing the field in the air gap and unbalanced force is compared to the results of a model based on conformal mapping [18] (the method frequently reported to provide accurate results) and 2D FEM. It is shown that the simplified model gives predictions of magnetic flux density in the air gap similar to the predictions of the conformal-mapping method and 2D FEM. Furthermore, the prediction of the force is, despite noticeable overestimation, useful and effective for machine-design purposes.

Modeling of machine losses with particular attention to their frequency dependence is given in the chapter. It is again demonstrated in the example of the test-machine laminations that classical representation of iron losses as a combination of eddy-current and hysteresis losses fails to account for the losses in actual laminated iron cores. Therefore, the modeling of iron losses relies on the manufacturer's data.

Using analytical models derived by Ferreira [23, 25], the dominant causes of copper losses in slotless PM machines are distinguished. It is shown that classical I^2R loss and eddy-current losses in the air-gap conductors have a dominant influence on the overall copper loss in a slotless machine while the skin- and proximity-effect influence can be neglected. Additionally, the section on copper losses adapts a Ferreira's equation to calculate eddy-current losses in the air-gap conductors of a slotless machine and correlates that expression to a simplified formula reported in literature (e.g. [26]). This study is one of the thesis' contributions.

Three different models of power of air-friction loss at the rotor cylindrical surface are combined with another model which represents the friction loss of a rotating disc. Since the author was not able to seriously examine validity of different models, these estimations represent a rough prediction of the total air-friction loss rather than a rigorous model.

An injection-molded plastic-bonded magnet is used in the test rotor; analytical models of eddy-current losses in rotor magnets can hardly be applied to such a magnet material. Nevertheless, given the material structure of plastic-bonded magnets (small magnetic particles in a plastic binder), it is expected that induced losses in the magnet are negligible.

References

1. Z. Zhu, D. Howe, E. Bolte, B. Ackermann, Instantaneous magnetic field distribution in brushless permanent magnet dc motors. i. open-circuit field. IEEE Trans. Magn. **29**(1), 124–135 (1993)
2. Z. Zhu, D. Howe, Instantaneous magnetic field distribution in brushless permanent magnet dc motors. ii. armature-reaction field. IEEE Trans. Magn. **29**(1), 136–142 (1993)
3. Z. Zhu, D. Howe, Instantaneous magnetic field distribution in brushless permanent magnet dc motors. iii. effect of stator slotting. IEEE Trans. Magn. **29**(1), 143–151 (1993)
4. Z. Zhu, D. Howe, Instantaneous magnetic field distribution in permanent magnet brushless dc motors. iv. magnetic field on load. IEEE Trans. Magn. **29**(1), 152–158 (1993)
5. Z. Zhu, K. Ng, N. Schofield, D. Howe, Improved analytical modelling of rotor eddy current loss in brushless machines equipped with surface-mounted permanent magnets. IEE Proc. Electr. Power Appl. **151**(6), 641–650 (2004)
6. H. Polinder, *On the Losses in a High-Speed Permanent-Magnet Generator with Rectifier*. Ph.D. Dissertation, Delft University of Technology, 1998
7. S. Holm, *Modelling and Optimization of a Permanent Magnet Machine in a Flywheel*. Ph.D. Dissertation, Delft University of Technology, 2003
8. H. Polinder, M. Hoeijmakers, Eddy-current losses in the segmented surface-mounted magnets of a pm machine. IEE Proc. Electr. Power Appl. **146**(3), 261–266 (1999)
9. H. Polinder, M. Hoeijmakers, M. Scuotto, Eddy-current losses in the solid back-iron of pm machines for different concentrated fractional pitch windings. in *Electric Machines Drives Conference, 2007. IEMDC '07. IEEE International*, vol. 1, pp. 652–657, May 2007
10. A. Borisavljevic, H. Polinder, J.A. Ferreira, Calculation of unbalanced magnetic force in slotless PM machines, in *Proceedings of Electrimacs*, 2011
11. C. Bi, Z. Liu, T. Low, Effects of unbalanced magnetic pull in spindle motors. IEEE Trans. Magn. **33**(5), 4080–4082 (1997)
12. Z. Liu, C. Bi, Q. Zhang, M. Jabbar, T. Low, Electromagnetic design for hard disk drive spindle motors with fluid film lubricated bearings. IEEE Trans. Magn. **32**(5), 3893–3895 (1996)
13. D. Guo, F. Chu, D. Chen, The unbalanced magnetic pull and its effects on vibration in a three-phase generator with eccentric rotor. J. Sound Vib. **254**(2), 297–312 (2002)
14. Z. Zhu, D. Ishak, D. Howe, C. Jintao, Unbalanced magnetic forces in permanent-magnet brushless machines with diametrically asymmetric phase windings. IEEE Trans. Ind. Appl. **43**(6), 1544–1553 (2007)
15. U. Kim, D. Lieu, Magnetic field calculation in permanent magnet motors with rotor eccentricity: without slotting effect. IEEE Trans. Magn. **34**(4), 2243–2252 (1998)
16. S. Swann, Effect of rotor eccentricity on the magnetic field in the air-gap of a non-salient-pole machine. Proc. Inst. Electr. Eng. **110**(11), 903–915 (1963)
17. J. Li, Z. Liu, L. Nay, Effect of radial magnetic forces in permanent magnet motors with rotor eccentricity. IEEE Trans. Magn. **43**(6), 2525–2527 (2007)
18. J.-P. Wang, D. Lieu, Magnetic lumped parameter modeling of rotor eccentricity in brushless permanent-magnet motors. IEEE Trans. Magn. **35**(5), 4226–4231 (1999)
19. F. Wang, L. Xu, Calculation and measurement of radial and axial forces for a bearingless pmdc motor, in *Industry Applications Conference, 2000. Conference Record of the 2000 IEEE*, vol. 1, pp. 249–252, 2000
20. Z. Zhu, K. Ng, D. Howe, Design and analysis of high-speed brushless permanent magnet motors, in *Electrical Machines and Drive, 1997 Eighth International Conference on (Conf. Publ. No. 444)*, pp. 381–385, 1–3 Sept 1997
21. A. Krings, J. Soulard, Overview and comparison of iron loss models for electrical machines. J. Electr. Eng. **10**(3), 162–169 (2010)
22. A. Borisavljevic, H. Polinder, J.A. Ferreira, Conductor optimization for slotless PM machines, in *Proceedings of the XV International Symposium on Electromagnetic Fields in Mechatronics, ISEF 2011*, 2011

23. J. Ferreira, Improved analytical modeling of conductive losses in magnetic components. IEEE Trans. Power Electron. **9**(1), 127–131 (1994)
24. J. Ferreira, Analytical computation of ac resistance of round and rectangular litz wire windings. IEE Proc. B Electr. Power Appl. **139**(1), 21–25 (1992)
25. J. Ferreira, *Electromagnetic Modeling of Power Electronic Converters* (Kluwer Academic Publishers, Norwell, 1989), Chap. 6
26. E. Spooner, B. Chalmers, âL˜TORUSâL™: a slotless, toroidal-stator, permanent-magnet generator. IEE Proc. B. Electr. Power Appl. **139**(6), 497–506 (1992)
27. J. Saari, *Thermal Analysis of High-Speed Induction Machines*. Ph.D. Dissertation, Acta Polytechnica Scandinavica, 1998
28. E. Bilgen, R. Boulos, Functional dependence of torque coefficient of coaxial cylinders on gap width and reynolds numbers. J. Fluids Eng. **95**(1), 122–126 (1973), http://link.aip.org/link/?JFG/95/122/1
29. Y. Yamada, Torque resistance of a flow between rotating co-axial cylinders having axial flow. Bull. JSME **5**(20), 634–642 (1962)
30. O. Aglen, A. Andersson, Thermal analysis of a high-speed generator, *Industry Applications Conference, 2003. 38th IAS Annual Meeting. Conference Record of the*, vol. 1, pp. 547–554, 12–16 Oct 2003
31. O. Aglen, Loss calculation and thermal analysis of a high-speed generator, in *Electric Machines and Drives Conference, 2003. IEMDC '03. IEEE International*, vol. 2, pp. 1117–1123, June 2003
32. O. Aglen, Back-to-back tests of a high-speed generator, in *Electric Machines and Drives Conference, 2003. IEMDC '03. IEEE International*, vol. 2, pp. 1084–1090, June 2003
33. L. Zheng, T. Wu, D. Acharya, K. Sundaram, J. Vaidya, L. Zhao, L. Zhou, K. Murty, C. Ham, N. Arakere, J. Kapat, L. Chow, Design of a super-high speed permanent magnet synchronous motor for cryogenic applications, in *Electric Machines and Drives, 2005 IEEE International Conference on*, pp. 874–881, 15 May 2005
34. M. Awad, W. Martin, Windage loss reduction study for tftr pulse generator, in *Fusion Engineering, 1997. 17th IEEE/NPSS Symposium*, vol. 2, pp. 1125–1128, Oct 1997
35. C. Zwyssig, S. Round, J. Kolar, Analytical and experimental investigation of a low torque, ultra-high speed drive system, in *Industry Applications Conference, 2006. 41st IAS Annual Meeting. Conference Record of the 2006 IEEE*, vol. 3, pp. 1507–1513, Oct 2006
36. J. Daily, R. Nece, Chamber dimenstion effects on induced flow and frictional resistance of enclosed rotating disks. ASME J. Basic Eng. **82**(1), 217–232 (1960)

Chapter 4
Structural Aspects of PM Rotors

4.1 Introduction

Structural design of a rotor of a high-speed electrical machine represents a challenging task. At high rotational frequencies centrifugal forces and, accordingly, stress in the rotor material become very high. Temperature increase in the rotor due to induced eddy currents and friction will produce additional, thermal stress between the rotor parts with different thermal properties. High-speed rotors must be capable of withstanding those stresses and also transfer of electromagnetic torque must be ensured over the whole rotor.

In a PM rotor, the magnet represents the most mechanically vulnerable part. While the compressive strength of permanent magnets is good, their flexural and tensile strengths are very low [1]. Magnets cannot sustain tension caused by centrifugal forces during high-speed rotation. Besides, magnets are usually very brittle and cannot be pressed onto the shaft. Therefore, a permanent magnet in a high-speed rotor must be contained in a non-magnetic enclosure or sleeve which would limit tension in the magnet and guarantee the transfer of torque from the magnet to the shaft at elevated speeds.

As a rule, the magnet in a high-speed rotor is either in the form of a full cylinder or consists of separated blocks that are glued on the steel shaft. Such a rotor structure is generally preferred for high-speed applications to a rotor with interior (buried) PMs since the latter is prone to much higher stress concentrations at high speeds [2]. The retaining sleeve is, most often, pressed on the magnet, although a more subtle technique is applying the sleeve on a cold-shrunken rotor.

Little has been published on mechanical design of high-speed PM rotors and optimization of the rotor structure. Still, wide literature on structural mechanics, including textbooks, provides sufficient information to analyze the subject. The thesis by Larsonneur [3] gives a general stress calculation for axisymmetrical rotors. He distinguished two main structural limitations for rotational speed: reaching yielding stress in one of the rotor parts or loss of contact between adjacent rotor parts. Larsonneur also observed the existence of an optimal interference fit between two

A. Borisavljević, *Limits, Modeling and Design of High-Speed Permanent Magnet Machines*, 71
Springer Theses, DOI: 10.1007/978-3-642-33457-3_4,
© Springer-Verlag Berlin Heidelberg 2013

rotating press- or shrink-fitted rings for maximum permissible speed using numerical solution for the stress in the rings. Binder et al. [2] showed advantages of using surface-mounted magnets for high speed and also the validity of analytical mechanical modeling for the case of magnets without inter-pole gaps. In choosing materials for high-speed generator, Zwyssig et al. [4] considered not only strength of rotor materials, but also compatibility of their thermal properties.

Aim of this chapter is to model the influence of rotational speed and mechanical fittings on stress in a high-speed rotor, while also considering the operating temperature. Through analytical modeling, structural limits for the rotor speed are determined and quantified. At the same time, a relatively simple approach of optimizing the rotor structure is achieved. This optimization approach will be revisited in Chap. 7 where design of a carbon fiber sleeve for the rotor of the test machine is presented.

Basic correlations between the structural quantities will not be elaborated in this chapter and the reader is referred to numerous textbooks on structural mechanics and elasticity.

The modeling of the stress first considers a single rotating cylinder and then a compound of cylinders, which is regarded as a good representative of high-speed PM rotors. The analytical models are then tested against finite element modeling for the example of the test-machine rotor. Finally, structural limits and the approach for the retaining sleeve design are presented.

4.2 Stress in a Rotating Cylinder

A model of a rotating hollow cylinder, whose cross-section is presented in Fig. 4.1, will be used in this section so as to form a basis for the structural analysis of PM rotors in the ensuing sections.[1] The cylinder rotates with a rotational speed Ω and is subjected to an internal static pressure p_i and an external static pressure p_o. Radial distribution of temperature increment $\tau = \tau(r)$ is known:

$$\tau = T - T_0, \tag{4.1}$$

where T_0 is the initial (room) temperature.

To facilitate a relatively simple analytical solution of the stress within the cylinder, an approach suitable for plane elastic problems will be taken. Figure 4.2 represents two types of cylinders whose stress or strain can be modeled planar. If the axial length of a cylinder is much smaller than its radius the cylinder can be modeled using the *plane stress* condition under which the stresses occur only in the plane perpendicular to the z-axis. Conversely, if the length of the cylinder is very large, *plane strain* condition under which, in general, axial strain is constant [6]. The plane strain condition is, thus, suitable for slender cylinders, thus for long rotors, while the plane *stress* condition is valid for disks, thus for short, disk-shaped rotors and laminated long rotors.

---

[1] Parts of this section have been taken from Borisavljevic et al. [5].

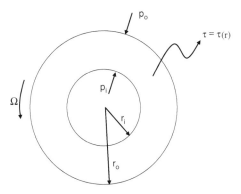

Fig. 4.1 Cross-section of a hollow rotating cylinder

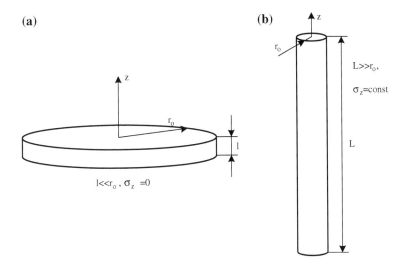

Fig. 4.2 Cylinders with plane **a** stress and **b** strain elastic conditions

4.2.1 Isotropic Modeling

In this subsection the material of the cylinder is assumed to be isotropic and linear. Hooke's law for the case of the axisymmetrical isotropic cylinder will be expressed in a generalized form to comprise both plane stress and strain conditions [6]:

$$
\underbrace{\begin{bmatrix} \varepsilon_r \\ \varepsilon_\theta \end{bmatrix}}_{\underline{\varepsilon}} = \underbrace{\begin{bmatrix} \dfrac{1}{E^*} & -\dfrac{v^*}{E^*} \\ -\dfrac{v^*}{E^*} & \dfrac{1}{E^*} \end{bmatrix}}_{\underline{M}} \underbrace{\begin{bmatrix} \sigma_r \\ \sigma_\theta \end{bmatrix}}_{\underline{\sigma}} + \underbrace{\begin{bmatrix} \alpha^* \\ \alpha^* \end{bmatrix}}_{\underline{\alpha}} \tau - \underbrace{\begin{bmatrix} c^* \\ c^* \end{bmatrix}}_{\underline{c}} \tag{4.2}
$$

where:

$$E^* = \begin{cases} \frac{E}{1-v^2}, & (*) \\ E, & (**) \end{cases} \qquad\qquad v^* = \begin{cases} \frac{v}{1-v}, & (*) \\ v, & (**) \end{cases}$$

$$\alpha^* = \begin{cases} (1+v)\alpha, & (*) \\ \alpha, & (**) \end{cases} \qquad\qquad c^* = \begin{cases} v\varepsilon_0, & (*) \\ 0, & (**) \end{cases} \qquad (4.3)$$

$(*)$—under the plane strain condition
$(**)$—under the plane stress condition

In (4.2) and (4.3) $\underline{\varepsilon}$ and $\underline{\sigma}$ denote 2D strain and stress, respectively, E is Young's modulus, v is Poisson's ratio and α is the coefficient of linear thermal expansion. The term ε_0 designates residual axial strain under the plane strain condition, i.e. $\varepsilon_z = \varepsilon_0$.

From (4.2) the vector of stress components is:

$$\underline{\sigma} = \underline{M}^{-1}\left(\underline{\varepsilon} + \underline{c}\right) - \underline{M}^{-1}\underline{\alpha} \cdot \tau \qquad (4.4)$$

To obtain the expression for stress distribution in the cylinder, the displacement technique will be employed. The correlations between radial displacement and strain are:

$$\varepsilon_r = \frac{du}{dr}, \quad \varepsilon_\theta = \frac{u}{r} \qquad (4.5)$$

and the force-equilibrium equation is:

$$r\frac{d\sigma_r}{dr} + \sigma_r - \sigma_\theta + \rho r^2 \Omega^2 = 0. \qquad (4.6)$$

Equations (4.5) and (4.6) are explained in more details in Appendix A.

By combining Eqs. (4.4), (4.5) and (4.6) we obtain the governing differential equation for radial displacement:

$$r^2 \frac{d^2u}{du^2} + r\frac{du}{dr} - u = -(1 - v^{*2})\frac{\rho r^3 \Omega^2}{E^*} + (1 + v^*)\alpha^* r^2 \frac{d\tau}{dr}, \qquad (4.7)$$

and its solution is given by:

$$u = Ar + \frac{B}{r} - (1 - v^{*2})\frac{\rho r^3 \Omega^2}{8E^*} + (1 + v^*)\alpha^* \frac{1}{r}\int_r \tau r\, dr. \qquad (4.8)$$

After substituting (4.8) into (4.5) and then into (4.4), expressions for the stress components yield:

$$\sigma_r = \frac{E^*}{1 - v^*}A - \frac{E^*}{1 + v^*}\frac{B}{r^2} - \frac{3 + v^*}{8}\rho r^2 \Omega^2 - \frac{\alpha^* E^*}{r^2}\int_r \tau r\, dr$$

$$\sigma_\theta = \frac{E^*}{1 - v^*}A + \frac{E^*}{1 + v^*}\frac{B}{r^2} - \frac{1 + 3v^*}{8}\rho r^2 \Omega^2 + \frac{\alpha^* E^*}{r^2}\int_r \tau r\, dr - \alpha^* E^* \tau \qquad (4.9)$$

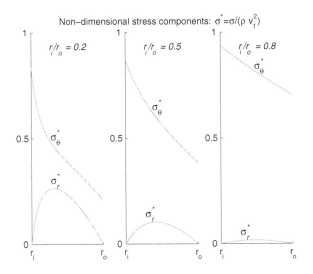

Fig. 4.3 Stress components resulting from the cylinder rotation

In the model of the rotating cylinder from Fig. 4.1 the boundary coefficients A and B are obtained by equating the radial stresses at the external cylinder surfaces with the static pressures:

$$\sigma_r\,(r_i) = -p_i, \ \sigma_r\,(r_o) = -p_o \tag{4.10}$$

After applying values for the boundary coefficients, stress components can be subdivided into parts based on rotation, static pressure and temperature increment:

$$\begin{aligned}
\sigma_r &= \sigma_r^{\Omega} + \sigma_r^{p} + \sigma_r^{\tau} \\
\sigma_\theta &= \sigma_\theta^{\Omega} + \sigma_\theta^{p} + \sigma_\theta^{\tau}
\end{aligned} \tag{4.11}$$

The rotation-influenced parts have the following values:

$$\begin{aligned}
\sigma_r^{\Omega} &= \frac{(3+v^*)\,\rho\Omega^2}{8}\left(r_o^2 + r_i^2 - \frac{r_o^2 r_i^2}{r^2} - r^2\right) \\
\sigma_\theta^{\Omega} &= \frac{(3+v^*)\,\rho\Omega^2}{8}\left(r_o^2 + r_i^2 + \frac{r_o^2 r_i^2}{r^2} - \frac{1+3v^*}{3+v^*}r^2\right)
\end{aligned} \tag{4.12}$$

The stress components are scaled down with the factor $1/(\rho v_t^2)$, where $v_t = \Omega r_o$ is the tip tangential speed, and they are plotted in Fig. 4.3 for $v^* = 0.3$. Evidently, the tangential stress is dominant throughout the whole cylinder's volume and its maximum occurs at the inner surface of the cylinder.

Fig. 4.4 Tresca's and von
Mises' yielding criteria in σ
plain

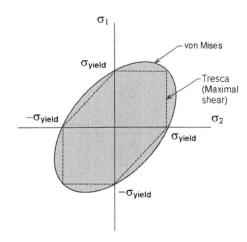

In order to avoid cracking, the value of stress in every point of an elastic body
needs to be kept lower than a certain value σ_U, which is either yield or ultimate stress,
depending on type of the material. Theories of material failure (Tresca's, von Mises':
see e.g. [7]), whose findings are confirmed with experimental results [8], define a
reference stress as a criterion for elastic failure. For the plane stress condition Tresca's
reference stress is defined as [3]:

$$\sigma_{ref}^{Tresca}(r) = \max\left(|\sigma_r(r) - \sigma_\theta(r)|, |\sigma_r(r)|, |\sigma_\theta(r)|\right),\tag{4.13}$$

and von Mises' reference stress as:

$$\sigma_{ref}^{vM}(r) = \sqrt{\sigma_r^2 + \sigma_\theta^2 - \sigma_r\sigma_\theta}.\tag{4.14}$$

Values of the reference stresses resulting from these theories are quite close, with
Tresca's criterion being slightly more conservative (see Fig. 4.4).

Reference stress must be lower than the yielding/ultimate stress of the material in
all points of the cylinder throughout the whole speed range:

$$\sigma_{ref}(r) < \sigma_U, \ r_i \le r \le r_o, \ \Omega \le \Omega_{\max}.\tag{4.15}$$

When considering the rotating cylinders it can be shown ([3], see also Fig. 4.3) that
the Tresca's reference stress is equal to the tangential stress, thus, with a maximum
value at the inner cylinder surface:

$$\sigma_{ref,\,\max} = \sigma_\theta(r_i) = \sigma_\theta^\Omega(r_i) + \sigma_\theta^{p,\,\tau}(r_i).\tag{4.16}$$

Only the rotation-dependent part of Eq. (4.16) will be analyzed:

$$\sigma_{ref,\max}^\Omega = \sigma_\theta^\Omega(r_i) = \frac{\rho\Omega^2}{4}\left[(3 + v^*)r_o^2 + (1 - v^*)r_i^2\right],\tag{4.17}$$

or, in a different form:

$$\sigma^{\Omega}_{ref,\max} = (3 + v^*)\frac{\rho v_t^2}{4} + (1 - v^*)\frac{\rho v_t^2}{4} \cdot \left(\frac{r_i}{r_o}\right)^2. \tag{4.18}$$

Apparently, the maximum reference stress in the cylinder is proportional to the square of the tip tangential speed, which can be used as an adequate limitation figure of speed of a rotating axisymmetrical body. Practically, from (4.18), since $v^* < 1$, the value of ρv_t^2 can be approximately taken as the maximum stress in a hollow rotating cylinder and that value must be below the material ultimate stress:

$$\sigma^{\Omega}_{ref,\max} \approx \rho v_t^2 < \sigma_U \tag{4.19}$$

In a lesser extent, the reference stress is also dependent on the ratio of the cylinder radii: the stress is lower for the smaller inner radius. However, the Eq. (4.18) cannot be used for calculation of the maximum stress in the example of a full cylinder ($r_i = 0$) because the Eqs. (4.9) and (4.12) are not defined for $r = 0$. In that case the boundary Eq. (4.10) change to:

$$u(0) \neq \infty, \ \sigma_r(r_o) = -p_o. \tag{4.20}$$

After calculating values of the boundary coefficients using (4.8), (4.9) and (4.20), the rotation-influenced parts of the rotor stress components yield:

$$\begin{aligned}
\sigma_r^{\Omega,full} &= \frac{(3 + v^*)\rho\Omega^2}{8}\left(r_o^2 - r^2\right) \\
\sigma_\theta^{\Omega,full} &= \frac{(3 + v^*)\rho\Omega^2}{8}\left(r_o^2 - \frac{1 + 3v^*}{3 + v^*}r^2\right)
\end{aligned} \tag{4.21}$$

Maximum reference stress of a full cylinder occurs in the center of the cylinder:

$$\sigma^{\Omega,full}_{ref,\max} = \sigma_\theta^{\Omega,full}(0) = (3 + v^*)\frac{\rho v_t^2}{8}. \tag{4.22}$$

The maximum stress value in this case can be approximated as:

$$\sigma^{\Omega,full}_{ref,\max} \approx \frac{\rho v_t^2}{2} < \sigma_U \tag{4.23}$$

Remarkably, the maximum reference stress of a full cylinder is equal to a half of the reference stress of a hollow cylinder of the same external radius with an infinitely small inner hole:

$$\sigma^{\Omega,full}_{ref,\max} = \frac{1}{2}\lim_{r_i \to 0^+}\sigma^{\Omega}_{ref,\max} \tag{4.24}$$

Consequently, the maximum achievable tangential speed of a full cylinder is, at least, $\sqrt{2}$ times higher than that of a hollow cylinder with the same external diameter.

4.2.2 Orthotropic Modeling

An assumption about isotropy of the cylinder's material was introduced in the beginning of this section. However, rotors of electrical machines sometimes include materials, such as fiber composites, with strong orthotropic nature, having, thus, quite different material properties in different directions. Analytical solution for stress in an orthotropic rotating cylinder is rather lengthy; therefore, only important correlations will be presented here which can, however, lead to desired expressions for the stress.

For the sake of simplicity, the temperature increment within the cylinder will be assumed uniform, thus:

$$\tau = \Delta T = const. \tag{4.25}$$

Generalized Hooke's law for plane stress/strain conditions in a hollow orthotropic cylinder has the following form:

$$\underbrace{\begin{bmatrix} \varepsilon_r \\ \varepsilon_\theta \end{bmatrix}}_{\underline{\varepsilon}} = \underbrace{\begin{bmatrix} \frac{1}{E_r^*} & -\frac{v_{r\theta}^*}{E_r^*} \\ -\frac{v_{\theta r}^*}{E_\theta^*} & \frac{1}{E_\theta^*} \end{bmatrix}}_{\underline{M}} \underbrace{\begin{bmatrix} \sigma_r \\ \sigma_\theta \end{bmatrix}}_{\underline{\sigma}} + \underbrace{\begin{bmatrix} \alpha_r^* \\ \alpha_\theta^* \end{bmatrix}}_{\underline{\alpha}} \tau - \underbrace{\begin{bmatrix} c_r^* \\ c_\theta^* \end{bmatrix}}_{\underline{c}}, \tag{4.26}$$

where:

$$E_r^* = \begin{cases} \frac{E_r}{1-v_{rz}v_{zr}}, & (*) \\ E_r, & (**) \end{cases} \qquad E_\theta^* = \begin{cases} \frac{E_\theta}{1-v_{\theta z}v_{z\theta}}, & (*) \\ E_\theta, & (**) \end{cases}$$

$$v_{r\theta}^* = \begin{cases} \frac{v_{r\theta}+v_{rz}v_{zr}}{1-v_{rz}v_{zr}}, & (*) \\ v_{r\theta}, & (**) \end{cases} \qquad v_{\theta r}^* = \begin{cases} \frac{v_{\theta r}+v_{\theta z}v_{z\theta}}{1-v_{\theta z}v_{z\theta}}, & (*) \\ v_{\theta r}, & (**) \end{cases}$$

$$\alpha_r^* = \begin{cases} \alpha_r + v_{rz}\frac{E_z}{E_r}\alpha_z, & (*) \\ \alpha_r, & (**) \end{cases} \qquad \alpha_\theta^* = \begin{cases} \alpha_\theta + v_{\theta z}\frac{E_z}{E_\theta}\alpha_z, & (*) \\ \alpha_\theta, & (**) \end{cases} \tag{4.27}$$

$$c_r^* = \begin{cases} v_{rz}\frac{E_z}{E_r}\varepsilon_0, & (*) \\ 0, & (**) \end{cases} \qquad c_\theta^* = \begin{cases} v_{\theta z}\frac{E_z}{E_\theta}\varepsilon_0, & (*) \\ 0, & (*) \end{cases}$$

(*)—under the plane strain condition
(**)—under the plane stress condition

(The reader may refer, for instance, to literature on mechanics of composite materials, e.g. [9].)

For an orthotropic material the following symmetry relation also holds:

$$\frac{v_{r\theta}}{E_r} = \frac{v_{\theta r}}{E_\theta}. \tag{4.28}$$

Similarly as in the beginning of this section, the Hooke's law in the stiffness form is obtained using (4.4). After combining the resulting expressions with (4.5) and (4.6), a differential equation over radial displacement yields the following form:

$$\frac{d^2 u}{dr^2} + \frac{1}{r}\frac{du}{dr} - h\frac{u}{r^2} + \frac{P}{r}\Delta T + Q\Omega^2 r = 0, \tag{4.29}$$

where h, P and Q are constants dependent on material properties:

$$h = \frac{E_\theta}{E_r}, \tag{4.30}$$

$$P = (1 - v_{r\theta})\alpha_\theta h - (1 - v_{\theta r})\alpha_r, \tag{4.31}$$

$$Q = \frac{(1 - v_{r\theta}v_{\theta r})\rho}{E_r}. \tag{4.32}$$

Solution of the differential Eq. (4.29) gives the expression for displacement:

$$u = Ar^{\sqrt{h}} + Br^{-\sqrt{h}} + \frac{P}{h-1}\Delta T r + \frac{Q}{h-9}\Omega^2 r^3, \tag{4.33}$$

and further derivation of the stress components can be performed in the same way as for the isotropic example.

It is noticeable that the analytical solving of the stress of an orthotropic body is rather burdensome. However, the orthotropic solution will provide a means for testing applicability of the isotropic assumption for structural modeling of a PM rotor.

4.3 Mechanical Stress in a PM Rotor

A PM rotor will be modeled as a compound of three adjacent cylinders which represent an (iron) shaft, permanent magnet and the magnet retaining sleeve (Fig. 4.5). Rotors of small high-speed PM generators consist sometimes, though, only of a solid magnet enclosed with a sleeve [10, 11] (i.e. $r_{Fe} = 0$): the configuration that renders lower stress in the magnet, as shown in the previous section. Still, conclusions from this and the subsequent chapter will remain valid for that rotor configuration too (See footnote 1).

In the rotor compound contact pressure between the adjacent cylinders must be maintained throughout the whole speed range. In other words, radial stress on the

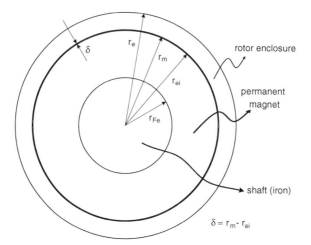

Fig. 4.5 Cross-section of the model of a PM rotor

boundary surfaces must be smaller than zero. In Sect. 4.2.1 (Eq. (4.12)) it was seen that radial stresses based on rotation at the boundaries is zero. Therefore a static pressure needs to be applied in order to maintain contact between the magnet and other rotor parts. The contact pressure is achieved by press- or shrink-fitting a non-magnetic retaining sleeve over the permanent magnet. The amount of shrinkage, or the interference fit, is defined as:

$$\delta = r_m - r_{ei}.\tag{4.34}$$

With help of the previous section, stress in the rotor will be correlated with the rotational speed, interference fit and operating temperature. The assumption (4.25) about a constant temperature increment in the rotor will be maintained for simplicity; having different temperatures $(\Delta T_{1,2,3})$ of different rotor parts would add, though, a little to complexity of the solution.

With the aforementioned assumption, regarding the parts of a fully isotropic PM rotor whose cross-section is given in Fig. 4.5, expressions (4.8) and (4.9) for displacement and stress components in rotating cylinders take on the following form:

$$u_i = A_i r + \frac{B_i}{r} - (1 - v^{*2})\frac{\rho r^3 \Omega^2}{8E^*}, \quad i = 1, 2, 3 \tag{4.35}$$

and

$$\sigma_{ri} = \frac{E^*}{1 - v^*}A_i - \frac{E^*}{1 + v^*}\frac{B_i}{r^2} - \frac{3 + v^*}{8}\rho r^2 \Omega^2$$

$$\sigma_{\theta i} = \frac{E^*}{1 - v^*}A_i + \frac{E^*}{1 + v^*}\frac{B_i}{r^2} - \frac{1 + 3v^*}{8}\rho r^2 \Omega^2 - \alpha^* E^* \Delta T$$

$$i = 1, 2, 3 \tag{4.36}$$

where 1, 2 and 3 denote cylindrical rotor components: the shaft, the magnet and the sleeve, respectively.

Boundary coefficients A_i and B_i in each region are obtained from the boundary equations which correlate radial stresses and displacements at the boundary regions:

$$
\begin{aligned}
&\sigma_{r1}(0) \neq \infty \\
&\sigma_{r2}(r_{Fe}) - \sigma_{r1}(r_{Fe}) = 0 \\
&u_{r2}(r_{Fe}) - u_{r1}(r_{Fe}) = 0 \\
&\sigma_{r3}(r_m) - \sigma_{r2}(r_m) = 0 \\
&u_{r3}(r_m) - u_{r2}(r_m) = \delta \\
&\sigma_{r3}(r_e) = 0
\end{aligned}
\tag{4.37}
$$

Solution of the system of Eq. (4.37) is achieved using a symbolic solver. Finally, after substituting obtained coefficients A_i and B_i into (4.36) analytical expressions for radial and tangential stress in the rotor are obtained. The expressions are rather lengthy; however, influences of static pressure (fitting), centrifugal force (rotation) and temperature increment on the stress can be clearly distinguished. Thus, both radial and tangential stresses can be expressed in the following way:

$$
\sigma_{r/\theta}(r) = \mathcal{F}_{r/\theta} \cdot \Omega^2 + \mathcal{G}_{r/\theta} \cdot \delta + \mathcal{H}_{r/\theta} \cdot \Delta T
\tag{4.38}
$$

where $\mathcal{F}$, $\mathcal{G}$ and $\mathcal{H}$ are functions of radius r, external cylinders' radii (r_{Fe}, r_m and r_e) and material properties (E^*, v^*, α^*).

Similarly, for those regions (cylinders) with orthotropic properties, Eqs. (4.35) and (4.36) for displacement and stress components can be replaced by corresponding expressions valid for the orthotropic cylinders. Again, solving system of the Eq. (4.37) will result in analytical expressions for stress components that retain the form given in (4.38).

In the remainder of this section validity of these models will be tested against 2D finite element (FE) modeling using the rotor of the test machine. The rotor will be referred to as the *test rotor*.

4.3.1 Test Rotor: Analytical Models

The test rotor has higher polar than transversal inertia having a disc which dominates its volume. A plastic-bonded magnet is applied onto the stainless steel disc in a ring shape. Finally, a carbon-fiber ring is pressed over the magnet with $\delta = 95\,\mu m$ of the press fit. The rotor was designed for operation at rotational speeds up to 200.000 rpm and temperatures up to 85 °C. Dimensions and properties of the parts of the test-rotor disc are given in Table 4.1.

In the table r_o denotes outer radius of the corresponding cylinder (r_{Fe}, r_m or r_e). The elastic moduli represent tensile moduli of the materials; it was assumed,

Table 4.1 Properties of the test-rotor parts

	r_o (mm)	ρ (g/cm³)	E_r	E_θ (GPa)	$\nu_{\theta r}$	ν_{rz}	α_r	α_θ (μm/m/°C)
Shaft: stainless steel	10.5	7.8	200		0.3		12	
Magnet: PPS-bonded NdFeB	14.5	4.8	31.7		0.3		4.7	
Sleeve: carbon fibers	16.5	1.6	9.5	186	0.3	0.59	−1	5

however, that the compressive and tensile moduli of each material are equal. Details on acquiring mechanical properties of the rotor materials are given in Chap. 7.

Calculations of mechanical stress in the rotor were performed based on two analytical models: one that assumes isotropic behavior of the carbon-fiber ring and other that takes into account full orthotropic data of the carbon fiber composite using the orthotropic cylinder modeling presented in Sect. 4.2.2. In the former model the carbon-fiber properties in the direction of fibers were assumed to be valid in all directions, thus: $E = E_\theta$, $\alpha = \alpha_\theta$ and $\nu = \nu_{\theta r}$. Since the test rotor has a disc shape, the plane stress condition was appropriate.

Both models resulted in components of stress in the form (4.38). For instance, after substitution of values of the mechanical properties from Table 4.1, the stress components in the magnet and retaining sleeve (carbon fibers) modeled with the isotropic assumption yield:

$$
\sigma_\theta^{iso}(r) =
\begin{cases}
\left(0.4 - \dfrac{1.2e-5}{r^2} - 1.1e3r^2\right) \cdot \Omega^2 + \left(-1.0e12 + \dfrac{4.8e7}{r^2}\right) \cdot \delta \\
\quad + \left(-8.3e4 + \dfrac{22}{r^2}\right) \cdot \Delta T \ (r_{Fe} < r \le r_m) \\[2ex]
\left(0.3 - \dfrac{3.4e-5}{r^2} - 370r^2\right) \cdot \Omega^2 + \left(4.3e12 + \dfrac{1.2e9}{r^2}\right) \cdot \delta \\
\quad + \left(6.4e5 + \dfrac{170}{r^2}\right) \cdot \Delta T \ (r_m < r \le r_e)
\end{cases}
\tag{4.39}
$$

$$
\sigma_r^{iso}(r) =
\begin{cases}
\left(0.4 + \dfrac{1.2e-5}{r^2} - 2.0e3r^2\right) \cdot \Omega^2 + \left(-1.0e12 - \dfrac{4.8e7}{r^2}\right) \cdot \delta \\
\quad + \left(-8.3e4 - \dfrac{22}{r^2}\right) \cdot \Delta T \ (r_{Fe} < r \le r_m) \\[2ex]
\left(0.3 - \dfrac{3.4e-5}{r^2} - 640r^2\right) \cdot \Omega^2 + \left(4.3e12 - \dfrac{1.2e9}{r^2}\right) \cdot \delta \\
\quad + \left(6.4e5 - \dfrac{170}{r^2}\right) \cdot \Delta T \ (r_m < r \le r_e)
\end{cases}
\tag{4.40}
$$

In Eqs. (4.39) and (4.40) values of the variables are in SI units.

Similar expressions are obtained by having orthotropic assumption for the carbon-fibers region.

Fig. 4.6 Cross-section of the
PM rotor—FE model

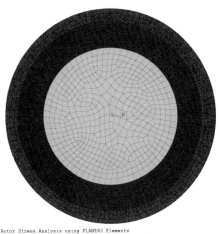

Rotor Stress Analysis using PLANE82 Elements

4.3.2 Test Rotor: 2D FE Model

Finite element modeling was performed using structural analysis of the Ansys FEM
software. The model used PLANE82 elements for 2D structural modeling while the
contact area between the magnet and enclosure was modeled using contact elements:
CONTA172 and TARGE169. Generated mesh is presented in Fig. 4.6.

4.3.3 Test Rotor: Results Comparison

Outcome of the three 2D models are presented in Table 4.2. The table represents
stresses at two critical regions: inner surfaces of the magnet and sleeve, including three examples: the rotor at standstill and the rotor rotating at the target speed
of 200.000 rpm at both room and maximum operating temperature. The index vM
denotes von Mises reference stress.

It can be seen that the isotropic model is, generally, in a good agreement with
other two modes. An exception is stress in the magnet at the maximum speed: when
the compression weakens, discrepancies between models become larger.

Nevertheless, as calculated by the FE model, compression in the magnet at standstill and tension in the fibres at the maximum speed is slightly smaller than predicted
by the isotropic model. Contact pressure between magnet and shaft is maintained
throughout the whole speed range. Temperature rise causes additional stress in the
rotor but also increases the contact pressure between the cylinders.

Although orthotropic analytical model predict the stress more accurately, the
accuracy improvement does not justify great increase in complexity with respect to
the isotropic model.

Table 4.2 Stress components at critical regions based on different analytical 2D models

Mechanical stress	20°C, 0 rpm			20°C, 200.000 rpm			85°C, 200.000 rpm		
[MPa]	Iso	Ortho	FEM	Iso	Ortho	FEM	Iso	Ortho	FEM
Sleeve, $\sigma_r(r_m)$	-118.3	-109.8	-109	-117	-108.7	-108	-128	-122.8	-117
Sleeve, $\sigma_\theta(r_m)$	921	940.6	935	1088	1108	1110	1176	1188	1160
Magnet, $\sigma_r(r_{Fe})$	-137.8	-128	-127	-26.8	-17.1	-13.1	-43.7	-37.4	-25.9
Magnet, $\sigma_\theta(r_{Fe})$	-56	-52	-51.6	-0.44	3.52	5.18	6.54	9.13	8.75
Magnet, $\sigma_{vM}(r_{Fe})$	120	111.3	110	26.6	19.1	16.3	47.3	42.7	31.2

In conclusion, the fully isotropic analytical model of mechanical stress is sufficiently accurate for modeling of PM rotors even when they include composite enclosures with strong orthotropic nature. The isotropic model will be used in this thesis for structural optimization of the test machine rotor. Design will be finally evaluated using 3D FE modeling in Chap. 7.

4.4 Structural Limits and Optimization of PM Rotors

In the preceding sections the influence of the geometry, operating speed, fitting and temperature on mechanical stress in a PM rotor was determined through analytical modeling. In this section, using the developed models, structural limits for the rotor speed will be determined and quantified. At the same time, a relatively simple approach of optimizing the rotor structure will be presented (see footnote 1).

In the example of the rotor consisting of cylindrical parts the structural design narrows down to defining thicknesses (or radii) of the rotor parts and the interference fits between them. However, in the case of a PM rotor, dimensions of the iron shaft and the magnet will strongly influence rotordynamic and electromagnetic performance of the machine. The structural design of the rotor can hardly be removed from the electromagnetic design of the whole machine. Here, it will be assumed that the dimensions of the shaft and magnet were defined beforehand during the electromagnetic design. Therefore, the optimization of the rotor structure will focus on determining thickness and fitting of the magnet retaining sleeve so as to maintain the structural integrity of the PM rotor throughout the whole ranges of operating speed and temperature.

The thickness of the sleeve ($r_e - r_m$) and the interference fit (δ) must be adequately chosen so that the contact pressure between the adjacent cylinders is always preserved, thus:

$$\sigma_r(r_{Fe}, r_m) < 0 \quad (0 \leq \Omega \leq \Omega_{nom},\ 0 \leq \Delta T \leq \Delta T_{max}). \tag{4.41}$$

At the same time, the reference stress in each rotor part must be considerably below the yielding/ultimate stress of the corresponding material:

$$\sigma_{ref}(r) < \sigma_U^i \quad (0 \leq r \leq r_e,\ 0 \leq \Omega \leq \Omega_{nom},\ 0 \leq \Delta T \leq \Delta T_{max}) \tag{4.42}$$

The purpose of the sleeve is to prevent high tension in the magnet and ensure the transfer of torque from the magnet to the shaft. The highest tension in the whole rotor occurs at the inner surface of the sleeve. Suitable materials for the magnet retainment have, thus, high tensile strength and low weight (glass and carbon composites, titanium).

At the same time, after shrink- or press-fitting the sleeve, the magnet is subjected to compression. Compressive strength of sintered magnets (which are regularly used in high-performance applications) is relatively high and, usually, compressive stress in the magnet does not represent a limitation for the fitting of the sleeve.

Hence, in this type of high-speed rotors (Fig. 4.5), the most critical stresses are radial (contact) stress at the magnet-iron boundary and tangential stress (tension) at the sleeve inner surface. As a result of modeling from the previous sections and after having the dimensions of the shaft and magnet specified, analytical expressions for these stresses can be presented as:

$$\sigma^m_{r,crit} = \sigma_r (r_{Fe}) = F_1 (r_e) \cdot \Omega^2 - G_1 (r_e) \cdot \delta - H_1 (r_e) \cdot \Delta T \tag{4.43}$$

$$\sigma^e_{\theta,crit} = \sigma^e_\theta (r_e) = F_2 (r_e) \cdot \Omega^2 + G_2 (r_e) \cdot \delta + H_2 (r_e) \cdot \Delta T \tag{4.44}$$

where $F_{1,2}$ and $G_{1,2}$ are positive functions of the sleeve outer radius r_e, $r_e > r_m$. Sign of the functions $H_{1,2}$ depends on the difference between the coefficients of thermal expansion of the magnet and the sleeve. For the sake of the analysis the functions $H_{1,2}$ will be also assumed positive (equivalent to $\alpha_e < \alpha_m$).

If the interference fit is very high, contact pressure between magnet and iron will be maintained ($\sigma^m_{r,crit} < 0$), but the maximum stress σ^e_U in the enclosure will be reached at a certain speed. Conversely, if the fit is very low, loss-of-contact limit ($\sigma^m_{r,crit} = 0$) will be met with increasing speed. It can be thus inferred from (4.43) and (4.44) that for an expected operating temperature there is an optimal value of the interference for which both limits defined by (4.41) and (4.42) are reached at a same rotational speed Ω_{max} (see Fig. 4.7 and [3]). This speed can be adjusted by the enclosure radius r_e so that the theoretical maximum rotational speed Ω_{max} is a considerable margin higher than the operating speed Ω_{nom}.

Hence, the optimal fit δ_{opt} and theoretical maximum speed Ω_{max} are obtained as functions of the sleeve radius r_e from the following system of equations:

$$\sigma^m_{r,crit} (\Omega = \Omega_{max}, \delta = \delta_{opt}, \Delta T = 0) = 0$$

$$\sigma^e_{\theta,crit} (\Omega = \Omega_{max}, \delta = \delta_{opt}, \Delta T = \Delta T_{max}) = \sigma^e_U \tag{4.45}$$

In reality, the higher the sleeve thickness is ($r_e - r_m$) the higher the maximum permissible rotational speed will be. On the other hand, that thickness is restricted by the available space in the machine air-gap, which is, as assumed, also decided on in the electromagnetic design.

However, structural and electromagnetic optimization can be carried out simultaneously. Namely, for defined dimensions of the permanent magnet there is an optimal value of the interference fit and a necessary sleeve thickness for reach-

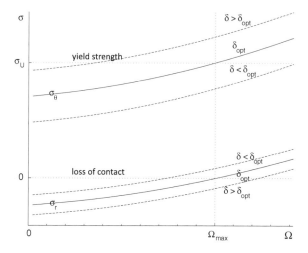

Fig. 4.7 Finding the optimal interference fit, similarly as in [3]

ing requested rotational speed. If this correlation is considered during design of the machine, requested electromagnetic performance (torque, power, losses) can be achieved with an optimal structural design of the rotor for the desired operating speed.

4.5 Conclusions

The permanent magnet represents the most mechanically vulnerable part of a PM rotor. In order to prevent high tension in the magnet and ensure the transfer of torque from the magnet to the shaft, high-speed PM rotors are usually enclosed with strong non-magnetic retaining sleeves. Good design of the retaining sleeve is crucial for the rotor structural integrity.

In the chapter, a high-speed rotor is represented as a compound of three concentric cylinders which represent the (iron) shaft, permanent magnet and magnet retaining sleeve. The chapter models the influence of rotational speed and mechanical fittings on stress in such a rotor, while also considering the operating temperature. The modeling of stress in a rotating PM rotor is based on equations which can be found in textbooks on structural mechanics. The suitability of 2D analytical models to represent a PM rotor without magnet-pole spacers is demonstrated by Binder in [2].

This thesis shows that a model which assumes isotropic behavior of the carbon-fiber retaining sleeve by assigning the properties in the direction of fibers to all directions makes almost equally good predictions as a fully orthotropic model of the rotor. Results of these two analytical models were compared to results of 2D FEM and agreement of the models is quite satisfactory.

The critical, thus limiting stresses in the rotor are radial (contact) stress at the magnet-iron boundary and tangential stress (tension) at the sleeve inner surface. It is shown in that for an expected operating temperature there is an optimal value of the interference fit between the sleeve and magnet for which both tension and contact limits are reached at an equal rotational speed. This speed can be adjusted by the enclosure thickness so that the theoretical maximum rotational speed is a considerable margin higher than the operating speed.

In the chapter, structural limits for speed of PM rotors are identified and the limiting parameters (stresses at the rotor material boundaries) are represented in a simple analytical form that clearly indicates optimal geometry of the rotor retaining sleeve. In this way, a relatively simple approach of optimizing the retaining sleeve is achieved; the approach takes into account the influence of rotational speed, mechanical fittings and operating temperature on stress in a high-speed rotor. This represents the main contribution of this chapter.

The presented structural design of the rotor follows the electromagnetic design of the machine. However, the analytical approach for the sleeve optimization lends itself to inclusion into a simultaneous, structural and electromagnetic optimization process. Namely, for defined dimensions of the permanent magnet there is an optimal value of the interference fit and a necessary sleeve thickness for reaching requested rotational speed. If this correlation is considered during the machine design, electromagnetic performance requirements (torque, power, losses) can be achieved with an optimal structural design of the rotor for the desired operating speed.

References

1. Y.M. Rabinovich, V.V. Sergeev, A.D. Maystrenko, V. Kulakovsky, S. Szymura, H. Bala, Physical and mechanical properties of sintered Nd-Fe-B type permanent magnets. Intermetallics **4**(8), 641–645 (1996)
2. A. Binder, T. Schneider, M. Klohr, Fixation of buried and surface-mounted magnets in high-speed permanent-magnet synchronous machines. IEEE Trans. Ind. Appl. **42**(4), 1031–1037 (2006)
3. R. Larsonneur, *Design and Control of Active Magnetic Bearing Systems for High Speed Rotation*. Ph.D. Dissertation, Swiss Federal Institute of Technology Zurich, 1990
4. C. Zwyssig, S.D. Round, J.W. Kolar, An ultra-high-speed, low power electrical drive system. IEEE Trans. Ind. Electron. **55**(2), 577–585 (2008)
5. A. Borisavljevic, H. Polinder, J.A. Ferreira, Enclosure design for a high-speed permanent magnet rotor, in *Proceeding of the Power Electronics, Machines and Drives Conference, PEMD 2010*, 2010
6. N. Naotake, B. Hetnarski Richard, T. Yoshinobu, *Thermal Stresses* (Taylor & Francis, New York, 2003), Chap. 6
7. A.C. Fischer-Cripps, *Introduction to Contact Mechanics* (Springer, Berlin, 2000)
8. R. LeMaster, *Steady Load Failure Theories—Comparison with Experimental Data*, Lectures on Machine Design, University of Tennessee at Martin, http://www.utm.edu/departments/engin/lemaster/machinedesign.htm
9. L. Kollar, G. Springer, *Mechanics of Composite Structures* (Cambridge University Press, Cambridge, 2003)

10. L. Zhao, C. Ham, L. Zheng, T. Wu, K. Sundaram, J. Kapat, L. Chow, C. Siemens, A highly efficient 200,000 rpm permanent magnet motor system. IEEE Trans. Magn. **43**(6), 2528–2530 (2007)
11. C. Zwyssig, J.W. Kolar, Design considerations and experimental results of a 100 W, 500,000 rpm electrical generator. J. Micromech. Microeng. **16**(9), 297–307 (2006)

Chapter 5
Rotordynamical Aspects of High-Speed Electrical Machines

5.1 Introduction

For very high-speed machines study of rotordynamics becomes very important. Rotors of today's high-speed machines regularly operate in supercritical regime, therefore, issues such as unbalance response and stability of rotation need to be addressed early in the design phase to prevent machine's failure.

Naturally, the field of rotordynamics concerns with a great number of rather complex problems, study of which goes well beyond scope of this thesis. Still, machine geometry, construction, even magnetic field have great influence on dynamical behavior of the rotor and that influence is particularly important when designing high-speed machines. In this chapter, a qualitative insight into important dynamical aspects of high-speed rotors will be given through analytical modeling. The goal of the chapter is to define the dynamical limits for the rotor speed and to correlate those limits with machine parameters.

The phenomenon which is of practical concern is rotor vibrations, their cause and influence on the system. Two types of vibrations can be distinguished: resonant and self-excited [1].

Resonant vibrations are excited by an oscillating force whose frequency coincides with one of the natural frequencies of the rotor-bearings system. Vibrations that are caused by an (external) oscillating force are generally referred to as *forced* vibrations.

Typically, the oscillating force comes from the rotor mass unbalance. Influence of the rotor unbalance (which is inevitably present in realistic rotors) can be modeled as a force which rotates around a perfectly balanced rotor and has the same rotational frequency as the actual rotor frequency. Therefore, the rotor unbalance will excite the resonant vibrations when rotational speed is equal to a natural frequency of the rotor-bearings system. Those rotational speeds are referred to as *critical speeds*.

Evidently, resonant vibrations occur at certain frequencies and are influenced by the amount of unbalance. They can be damped by external damping and may be passed if sufficient energy (dissipated in bearings' dampers) is invested.

However, not only mass unbalance of the rotor can cause resonant vibrations. Circulating fluids in the bearings or harmonics of unbalanced magnetic pull of the

A. Borisavljević, *Limits, Modeling and Design of High-Speed Permanent Magnet Machines*, Springer Theses, DOI: 10.1007/978-3-642-33457-3_5,

electrical machine whose forces may not be synchronous with the rotor can also excite resonant vibrations. Nevertheless, these phenomena will not be analyzed in detail in this chapter and the reader will be referred to literature.

Self-excited vibrations, on the other hand, require no external force for inception[1] [1, 2]. They arise within range(s) of rotational speed, usually after a certain (threshold) speed which is correlated with intrinsic properties of the system. The external (non-rotating) damping can increase threshold speed value but has virtually no influence when the vibrations occur. Unbalance has no influence on these vibrations—they would also occur in a perfectly balanced rotor.

Self-excited vibrations are unstable and hazardous; the rotor should not operate in the speed range where these vibrations occur. Hence, the threshold speed is the definite speed limit of the rotor. A rotor operating at speed higher than the threshold speed can be viewed as is in the state of unstable equilibrium—any perturbation will bring about vibrations whose amplitude will (theoretically) grow to infinity.

This chapter starts by defining different vibrational modes. A theoretical study on stability of rotation follows that will assess threshold speed of self-excited vibrations and, consequently, define practical speed limits of electrical machines with respect to critical speeds and properties of the bearings. The study lastly gives basic correlations between critical speeds and the rotor and bearings parameters. In the rest of the chapter only rigid rotors are considered. Unbalance response of rigid rotors is modeled analytically and suitability of different rotor geometries for high-speed rotation is analyzed.

The analysis in this chapter had a defining influence on the new high-speed-spindle concept presented in Chap. 7. The new spindle design has its foundation in conclusions drawn in the current chapter: the design is developed so that the major rotordynamical stability limits are simply avoided.

The list of literature that is relevant for the subject of this chapter is quite long; this study was mainly influenced by the works of Genta [1, 3, 4], Childs [2, 5] and Muszynska [6, 7].

5.2 Vibration Modes

In literature on electrical machines rigid and flexural (bending) vibration modes are often distinguished based on whether the rotor is deformed while vibrating. However, this division is rather conditional. Truly rigid modes can occur only in the case of free rotors. Otherwise, the modes are mixed. In practice, if the rotor stiffness is much higher than the stiffness of bearings (*soft-mounted* rotor), term *rigid-body modes* refers to the modes in which rotor deformation is negligible when compared to the

---

[1] The term *self-excited* can be disputed since a rotor that resonates due to its own unbalance is, physically, self-excited. However, by *self-excitation* is emphasized that no disturbing force is required for the vibration. On the other hand, the force coming from the rotor unbalance is considered external to the *ideal* rotor.

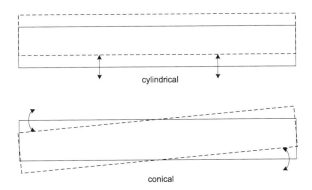

Fig. 5.1 Rigid-body vibration modes

Fig. 5.2 The first flexural
vibration mode

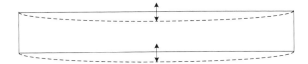

bearing deformation. If the condition for stiffness is not fulfilled, it is not possible to speak about rigid modes (Figs. 5.1 and 5.2).

Rigid-body vibrations occur at relatively low speeds. Since shape of the rotor is maintained internal rotor damping has no influence on the dynamics of the system.

Regarding the relative motion of the rotor with respect to geometrical axis cylindrical and conical modes are recognized.

Flexural (or bending) modes are higher order vibrations in which both rotor and bearings are deformed, therefore internal damping of rotor also plays role in the rotor dynamics. Flexural vibrations occur at higher speeds with respect to rigid vibrations.

Finally, torsional and axial vibrations can also occur, however, if flexural and torsional/axial modes are not coupled radial forces (e.g. unbalance) will not excite these vibrations.

5.3 Threshold of Instability

Purpose of this section is to describe mechanisms that lie behind rotational instability of high-speed machines. Through simplified analytical models of the rotor-bearing systems basic expressions for limiting rotational speed will be given and, more importantly, implications for high-speed electrical machines will be drawn.[2]

To find the threshold of instability the Jeffcott rotor model will be used. It is rather simplified model in which a rotor is represented as a massless (generally compliant)

---

[2] Parts of this section have been taken from Borisavljevic et al. [8] ©2010 IEEE.

Fig. 5.3 Balanced Jeffcott
rotor ©2010 IEEE

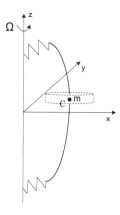

shaft with the mass concentrated at one point of the shaft—Fig. 5.3. However, this
model gives a qualitative insight into rotors' dynamical behavior and will be used
here to infer about the stability limit in general [1, 5].

5.3.1 Stability of the Jeffcott Rotor with Damping

For calculating the instability threshold the example of a perfectly balanced Jeffcott
rotor is considered, thus, center of rotor inertia coincides with the geometrical center
of the rotor C. The center is always located in the xy plane.

The equations of motion of the Jeffcott rotor in the stationary reference frame are:

$$m\frac{d^2 x_C}{dt^2} + c_n\frac{dx_C}{dt} + c_r\left(\frac{dx_C}{dt} + \Omega y_C\right) + kx_C = F_x$$

$$m\frac{d^2 y_C}{dt^2} + c_n\frac{dy_C}{dt} + c_r\left(\frac{dy_C}{dt} - \Omega x_C\right) + ky_C = F_y$$

(5.1)

or in the matrix form:

$$\begin{bmatrix} m & 0 \\ 0 & m \end{bmatrix}\begin{bmatrix} \ddot{x}_C \\ \ddot{y}_C \end{bmatrix} + \begin{bmatrix} c_n + c_r & 0 \\ 0 & c_n + c_r \end{bmatrix}\begin{bmatrix} \dot{x}_C \\ \dot{y}_C \end{bmatrix} + \left(\begin{bmatrix} k & 0 \\ 0 & k \end{bmatrix} + \Omega\begin{bmatrix} 0 & c_r \\ -c_r & 0 \end{bmatrix}\right)\begin{bmatrix} x_C \\ y_C \end{bmatrix} = \begin{bmatrix} F_x \\ F_y \end{bmatrix}$$

(5.2)

In (5.2) Ω is rotational speed, m the rotor mass, k stiffness of the bearings,
c_n is coefficient of viscous-type non-rotational (external) damping of the rotor-
bearings system and c_r is coefficient of viscous-type rotational (usually rotor internal)
damping.

Position of the point C can be also expressed as a complex vector:

$$r_C = x_C + jy_C = r_{Co}e^{st}. \tag{5.3}$$

Exponential form of r_C from (5.3) can be applied to (5.2) and the free whirling case ($F_x = F_y = 0$) yields characteristic equation of the system:

$$ms^2 + (c_r + c_n)s + k - j\Omega c_r = 0. \tag{5.4}$$

Solutions of Eq. (5.4) yield eigenvalues of the system:

$$s = \sigma + j\omega = -\frac{c_r + c_n}{2m} \pm \sqrt{\frac{(c_r + c_n)^2 - 4m(k - j\Omega c_r)}{4m^2}}. \tag{5.5}$$

Evidently, there are two complex eigenvalues:

$$s_{1,2} = \sigma_{1,2} + j\omega_{1,2}. \tag{5.6}$$

Imaginary part of the solutions represents frequency of the rotor free whirl and from (5.5) it yields the following form:

$$\omega^*_{1,2} = \pm\sqrt{\sqrt{\Pi^{*2} + \Omega^{*2}\zeta_r^2} + \Pi^*} \tag{5.7}$$

where:

$$\Pi^* = \left[1 - (\zeta_n + \zeta_r)^2\right]/2 \tag{5.8}$$

and ω^*, Ω^*, ζ_n and ζ_r are non-dimensional values of frequencies and dampings:

$$\omega^* = \frac{\omega}{\sqrt{k/m}} \quad \Omega^* = \frac{\Omega}{\sqrt{k/m}}$$
$$\zeta_n = \frac{c_n}{2\sqrt{km}} \quad \zeta_r = \frac{c_r}{2\sqrt{km}} \tag{5.9}$$

The solution of the characteristic Eq. (5.4) which has positive natural frequency (ω_1) represents the *forward* rotor whirl, which is in the direction of rotation, and the other solution with (ω_2) represents the *backward* whirl. In the case of an undamped rotor ($\zeta_n = \zeta_r = 0$) frequencies of the resonant whirls yield:

$$\omega_{1,2} = \pm\sqrt{\frac{k}{m}} = \pm\Omega_{cr} \tag{5.10}$$

where $\Omega_{cr} = \sqrt{k/m}$ denotes critical speed of the undamped Jeffcott rotor.

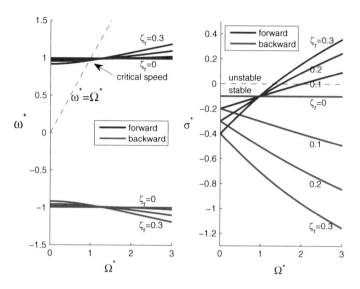

Fig. 5.4 Non-dimensional parts of complex eigenvalues of the Jeffcot rotor for different values of rotating damping and $\zeta_n = 0.1$

Real parts of the eigenvalues (5.5) are given by:

$$\sigma^*_{1,2} = -(\zeta_n + \zeta_r) \pm \sqrt{\sqrt{\Pi^{*2} + \Omega^{*2}\zeta_r^2} - \Pi^*}, \qquad (5.11)$$

where:

$$\sigma^* = \frac{\sigma}{\sqrt{k/m}} \qquad (5.12)$$

In order for the rotation to be stable it is necessary that both complex solutions (5.6) of (5.4) have negative real parts, thus $\sigma_{1,2} < 0$. It is apparent from (5.11) that the backward whirl is always stable ($\sigma_2 < 0$) while it is easy to show that the condition for stability of the forward whirl ($\sigma_1 < 0$) is equivalent to:

$$\Omega < \sqrt{\frac{k}{m}} \left(1 + \frac{c_n}{c_r}\right), \qquad (5.13)$$

or, in another form:

$$\Omega < \Omega_{cr} \left(1 + \frac{c_n}{c_r}\right). \qquad (5.14)$$

Vibrations (whirls) defined by (5.3) and (5.6) are essentially self-excited since the rotor is assumed perfectly balanced and no external force is acting. If the condition (5.14) is fulfilled these vibrations are hardly noticeable—the motions damp out very quickly. However, if the rotational speed is higher than the limit defined by (5.14) the amplitude of the vibrations will grow and will be limited only by the system non-linearities [1]. The rotor-bearings system from Fig. 5.3 is unstable in that case regardless of the presence of exciting forces such as mass unbalance or unbalanced magnetic force.

From (5.14) it is evident that the presence of internal or rotating damping can negatively influence the stability of rotation in the supercritical regime. If no rotating damping is present the system is always stable; when rotating damping is increased the instability threshold lowers (Fig. 5.4). On the other hand, non-rotating damping has always stabilizing effect [1, 5].

An insightful demonstration of the destabilizing effect of the rotating damping in the supercritical regime is presented in [9]. Similarly to an example from the paper, an intuitive explanation of the rotor instability is given in Appendix B.

The aforementioned analysis was performed for a simplified rotor model and the dampings are assumed to be of viscous type which is usually applicable only to the rotating damping. Still, important conclusions resulting from expression (5.14) also hold true when more complex models are used. More rigorous mathematical modeling was presented for instance in [10] using the Timoshenko beam model with incorporated viscous and hysteretic damping. The modeling showed that the internal viscous damping has stabilizing effect on the rotor stability in the subcritical regime but destabilizing effect in the supercritical regime. It was pointed out, however, that the hysteretic damping is also destabilizing in the subcritical regime. The influence of internal and external damping on the presence of self-excited vibrations was practically shown, for instance, in [11].

In the light of the forgoing modeling some important conclusions on rotordynamical stability of electrical machines will be given, as follows.

Rotation can become unstable in the supercritical regime of a certain vibration mode if rotating damping is present in that mode. The threshold speed can be increased with additional external damping.

During rigid-body vibrations, no deformation of rotor occurs and the internal rotor damping plays no role. Rotors are usually stable in supercritical range pertaining to rigid modes and today's high-speed rotors operate regularly in that speed range. An exception is rotors which are supported in fluid journal bearings—as result of non-synchronously circulating fluid, rotating damping influences also rigid-body modes and a rotor can become unstable beyond the rigid critical speed(s) [6, 7, 12, 13] (see the Sect. 5.3.2).

Rotors which possess some internal damping, can easily become unstable in supercritical range pertaining to flexural modes. Sources of rotating damping [2, 14] in electrical machines in general are eddy-currents [9], press fits, material damping, interaction with fluids [15] etc. Moreover, flexural modes are often poorly damped externally—bearings are located at the ends of the rotor and/or have insufficient bandwidth to cope with those vibrations.

Rotors of electrical machines are receptive to eddy-currents, always comprised of fitted elements, and often contain materials, such as composites, with significant material damping. Therefore, these rotors are prone to be unstable in flexural supercritical regime. The first flexural critical speed practically represents rotordy-namical speed limit of an electrical machine.

5.3.2 Jeffcott Rotor with Non-Synchronous Damping

It is of practical interest to study the case of the rotor-bearings system in which an energy dissipating element (damper) exists which rotates with a speed other than the speed of the rotor. A typical example for such systems are rotors in lubricated journal bearings or hydrodynamic bearings [16]. The average tangential speed of the bearing fluid differs, as a rule, from the tangential speed of the rotor giving a way to non-synchronous damping forces.

The equation of motion of the balanced Jeffcott rotor in presence of non-synchronous damping is as follows [3] (see [17] for gyroscopic rotors):

$$m\ddot{r}_C + c\dot{r}_C + (k - i\Omega c_r - i\Omega_d c_d)r_C = 0, \tag{5.15}$$

where Ω_d is rotational speed of the non-synchronously rotating damper, c_d is equivalent viscous damping correlated with that damper and $c = c_n + c_r + c_d$.

After seeking a solution in the form (5.3), the characteristic equation becomes:

$$ms^2 + cs + k - i\Omega c_r - i\Omega_d c_d = 0. \tag{5.16}$$

Solving the characteristic equation results in complex eigenvalues $s_{1,2} = \sigma_{1,2} + i\omega_{1,2}$ whose real parts in the non-dimensional form (5.12) are given by [3]:

$$\sigma^* = -\zeta \pm \sqrt{-\Pi^* + \sqrt{\Pi^{*2} + \Xi^{*2}}}, \tag{5.17}$$

where:

$$\Xi^* = \Omega^* \zeta_r + \Omega_d^* \zeta_d \tag{5.18}$$

and:

$$\Pi^* = \frac{1 - \zeta^2}{2}, \ \zeta = \frac{c}{2\sqrt{km}}. \tag{5.19}$$

It can be shown that the condition for rotation stability is equivalent to:

$$\Omega^* < \frac{\zeta}{\zeta_r} - \Omega_d^* \frac{\zeta_d}{\zeta_r} \quad \text{for } \Xi^* > 0, \tag{5.20}$$

$$\Omega^* < -\frac{\zeta}{\zeta_r} - \Omega_d^* \frac{\zeta_d}{\zeta_r} \quad \text{for } \Xi^* < 0. \tag{5.21}$$

If the spinning speed of the rotating damping is proportional to the rotor speed, it is reasonable to introduce the ratio between those two speeds $\alpha = \Omega_d / \Omega$; the stability condition can be then expressed as[3]:

$$\Omega^* < \frac{\zeta}{\zeta_r + \alpha \zeta_d} \quad \text{for } \Xi^* > 0, \tag{5.22}$$

$$\Omega^* < -\frac{\zeta}{\zeta_r + \alpha \zeta_d} \quad \text{for } \Xi^* < 0. \tag{5.23}$$

It is interesting to analyze the example of a rigid rotor. In that case, rotating damping has no influence on the system and, if the damper rotates in the same direction as the rotor, the condition of stability becomes:

$$\Omega < \frac{\Omega_{cr}}{\alpha} \left(1 + \frac{c_n}{c_d}\right) \tag{5.24}$$

where Ω_{cr} is the critical speed of the undamped Jeffcott rotor.

Finally, if the rotating damper has the dominant influence on the rotor damping, then $c_d \gg c_n$ and the condition for stability is approximately:

$$\Omega < \frac{\Omega_{cr}}{\alpha}. \tag{5.25}$$

In fluid journal bearings value of the ratio α is near 0.5 [3, 6, 7, 13, 15, 16] and the rotation becomes unstable at the speed which is approximately twice of the first critical speed. Subsynchronous rotor vibrations that exist due to subsynchronous spinning of the damper become unstable when their frequency reaches critical frequency. The phenomenon is known as *oil whirl* and at the threshold rotational speed defined by (5.25) it is actually replaced by the *oil whip*, a particularly hazardous form of instability in journal bearings [6, 7, 13] (see the Chap. 6 for further discussion on this issue).

Nature of these phenomena is very complex [6], explanation of which exceeds the outreach of this study. However, it is important to notice that rigid rotors can also become unstable in the presence of non-synchronous damping. In the study in the previous subsection it was shown that rotors supported by purely stationary bearings cannot become unstable unless rotor deforms and rotating damping becomes active. Here, an additional mechanism of rotational instability is described which is connected solely to the rotor-bearings interaction.

---

[3] From (5.22) and (5.23) it is evident that for $\alpha = -\zeta_r / \zeta_d$ (counter-rotating damping) the system is always stable, which can be useful in design of active dampers.

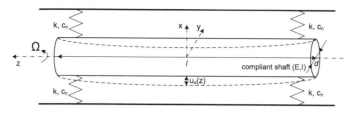

Fig. 5.5 A compliant rotating beam with supports

5.4 Critical Speeds Calculation

The goal of this section is to correlate values of critical speeds, in particular of
those connected with bending modes, to machine parameters. In order to calculate
critical speeds of a typical rotor-bearings system, the model of an axisymmetrical
Timoshenko's beam is going to be used. The system with a rotating beam is depicted
in Fig. 5.5 (See footnote 2).

The Timoshenko's beam is a comprehensive model of a continuous shaft that takes
into account shear deformation and gyroscopic coupling between dynamic behavior
of the shaft in different directions.

Displacement of the shaft $u = u_x + j u_y$ can be represented as:

$$u\,(z, t) = q\,(z)\, e^{j\omega t}, \tag{5.26}$$

where $q(z)$—*mode shapes*—must comply with:

$$EI_y \frac{d^4 q\,(z)}{dz^4} + \rho I_y \left[\omega^2 \left(1 + \frac{E\chi}{G} \right) - 2\omega\Omega \right] \frac{d^2 q\,(z)}{dz^2}$$
$$+ \rho \left[\omega^4 \frac{\rho I_y \chi}{G} - 2 \frac{\rho I_y \chi}{G} \omega^3 \Omega - A\omega \right] q\,(z) = 0 \tag{5.27}$$

(see e.g. [1] for details).

In (5.27) $I_y = I_x = d^4 \pi / 64$ is the surface moment of inertia of the shaft cross-
section, and χ is the shear parameter that has the value of $10/9$ for circular cross-
sections. Besides, ρ is the mass density, E is Young's modulus and G is the shear
modulus and it holds:

$$\frac{E}{G} = 2\,(1 + \nu), \tag{5.28}$$

where ν is Poisson's ratio.

From the boundary conditions determined by how the rotor is borne— free, supported, clamped, etc.—characteristic equations are obtained from which critical frequencies can be calculated.

5.4.1 Hard-Mounted Shaft

For a hard-mounted rotor ($k \rightarrow \infty$) supported at the rotor ends boundary conditions yield:

$$q\,(0) = q\,(l) = 0$$
$$\left.\frac{d^2 q}{dz^2}\right|_{z=0} = \left.\frac{d^2 q}{dz^2}\right|_{z=l} = 0 \tag{5.29}$$

and it can be shown that, as result of (5.29) $q\,(z)$ has the following form:

$$q\,(z) = q_{i_0} \sin\,(i \pi z / l)\,. \tag{5.30}$$

After substituting (5.30) into (5.27) the characteristic equation of the motion is obtained:

$$\omega^{*4} - 2\Omega^* \omega^{*3} - 4\omega^{*2}\frac{i^2 \lambda^2}{\pi^2 \chi^*}\left(1 + \chi^* + \frac{4\lambda^2}{i^2 \pi^2}\right) + 8\Omega^* \omega^* \frac{i^2 \lambda^2}{\pi^2 \chi^*} + 16\frac{i^4 \lambda^4}{\pi^4 \chi^*} = 0. \tag{5.31}$$

In (5.31) $\lambda = l/d$ is the rotor slenderness while ω^*, Ω^* and χ^* are non-dimensional natural frequency, rotational frequency and shear parameter of the rotor, respectively:

$$\omega^* = \frac{\omega}{\frac{\pi^2}{l^2}\sqrt{\frac{EI_y}{\rho A}}}, \quad \Omega^* = \frac{\Omega}{\frac{\pi^2}{l^2}\sqrt{\frac{EI_y}{\rho A}}}, \quad \chi^* = \frac{\chi E}{G} \tag{5.32}$$

Equation (5.31) implicitly defines correlation between natural frequencies of the rotor-bearings system and rotational speed. When plotted, lines $\omega^*\,(\Omega^*)$ are referred to as a *Campbell diagram* of the system and critical speeds represent intersections between those lines and line $\omega^* = \Omega^*$ (Fig. 5.6). Hence, from (5.31) and $\omega^* = \Omega^* = \Omega^*_{cr}$ the characteristic equation yields:

$$\Omega^{*4}_{cr} + 4\Omega^{*2}_{cr}\frac{i^2 \lambda^2}{\pi^2 \chi^*}\left(-1 + \chi^* + \frac{4\lambda^2}{i^2 \pi^2}\right) - 16\frac{i^4 \lambda^4}{\pi^4 \chi^*} = 0. \tag{5.33}$$

For $i = 0$ no real solutions for Ω^*_{cr} exist thus no rigid modes are present (refer to Eq. (5.30)). For $i > 0$ solutions of (5.33) are given by:

$$\Omega^*_{cr} = \frac{2i\lambda}{\pi}\sqrt{\frac{\left(-1 + \chi^* + \frac{4\lambda^2}{i^2 \pi^2}\right)\left(-1 + \sqrt{1 + \frac{4\chi^*}{\left(-1 + \chi^* + \frac{4\lambda^2}{i^2 \pi^2}\right)^2}}\right)}{2\chi^*}}. \tag{5.34}$$

Figure 5.7 shows non-dimensional values of first four critical speeds for different slenderness with $\chi = 10/9$ and $\nu = 0.3$. For very slender hard-mounted beams non-dimensional critical speeds approach to the square of the order i, namely:

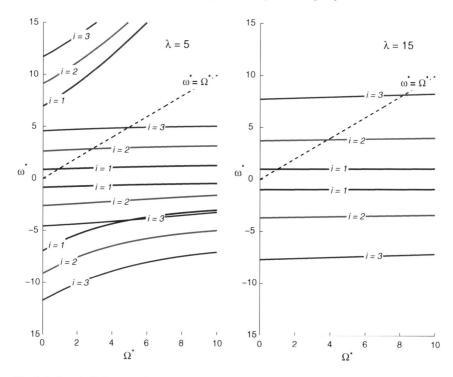

Fig. 5.6 Campbell diagram of a hard-mounted shaft based on Eq. 5.31; plotted for first three orders of critical speeds and two values of slenderness

$$\Omega^*_{cr}\big|_{\lambda\to\infty} = i^2, \quad i \geq 1, \tag{5.35}$$

the result following also from the Euler-Bernoulli beam model [1] which neglects influences of gyroscopic effect and shear deformation.

All the critical speeds in this case are connected with flexural vibrational modes. From (5.32) and (5.35) value of the first flexural critical speed in the example of a hard-mounted shaft can be approximated as:

$$\Omega_{cr} \approx \frac{\pi^2}{l^2}\sqrt{\frac{E I_y}{\rho A}} = \frac{\pi^2 d}{4 l^2}\sqrt{\frac{E}{\rho}}. \tag{5.36}$$

5.4.2 General Case: Bearings with a Finite Stiffness

High-speed rotors are predominantly soft-mounted, meaning that the stiffness of their bearings is by far lower than the rotor internal stiffness. For a soft-mounted rotor finite stiffness k of the bearings must be taken into account when calculating values of the critical speeds.

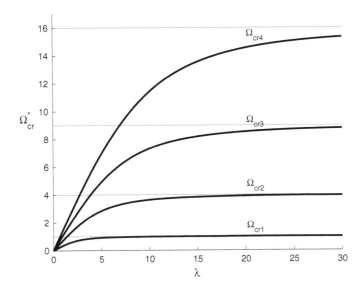

Fig. 5.7 First four critical speeds of a hard-mounted beam with respect to slenderness

Unbalanced magnetic force in a machine can also have noticeable impact on values of critical speeds. Resultant attraction force between the stator and the rotor appears as a *negative stiffness* of the machine, causing critical speeds to decrease. However, the influence of the machine stiffness will be neglected in this section and the reader is referred to literature on that subject (e.g. [18]).

To model the bearings stiffness in general and also to infer about vibration modes at calculated critical speeds, non-dimensional stiffness will be used:

$$k^* = \frac{kl^3}{EI_y} \tag{5.37}$$

Non-dimensional stiffness k^* practically represents ratio between the bearings stiffness and flexural stiffness of the beam (e.g. flexural stiffness of a simply supported beam is $48EI_y/l^3$). If k^* is very small ($k^* < 1$), first two critical speeds (or only first in the example of a disc, see Sect. 5.5) may be considered rigid. Otherwise, it cannot be spoken about rigid nor flexural critical speeds—the modes are mixed. Finally, when $k^* \gg 1$ the modeling from the previous subsection is applicable: practically, only flexural vibrations exist.

General solution of the differential Eq. (5.27) has the following form:

$$q(z) = C_1 e^{az/l} + C_2 e^{-az/l} + C_3 e^{jbz/l} + C_4 e^{-jbz/l}, \tag{5.38}$$

where $\pm a/l$ and $\pm jb/l$ are roots of the polynomial corresponding to Eq. (5.27).

Characteristic equation is obtained from boundary conditions. For a shaft supported at the ends shear force at the coordinates of the bearings must be equal to the bearings reaction:

$$\frac{d^3 q\,(z)}{dz^3}\bigg|_{z=0,l} = \frac{k^*}{l^3} q\,(z)\bigg|_{z=0,l}. \tag{5.39}$$

Condition of free rotation yields the second boundary condition—bending moment vanishes at the bearings coordinates:

$$\frac{d^2 q\,(z)}{dz^2}\bigg|_{z=0,l} = 0. \tag{5.40}$$

After substituting (5.38) into the boundary conditions (5.39) and (5.40) and setting $\omega^* = \Omega^* = \Omega^*_{cr}$ a four equations system is obtained which can be presented in a matrix form:

$$[A] \cdot [C] = 0, \tag{5.41}$$

where:

$$[A] = \mathcal{F}(\Omega^*_{cr}), \tag{5.42}$$

and:

$$[C] = [C_1\ C_2\ C_3\ C_4]^T. \tag{5.43}$$

The system (5.41) has a non-trivial solution $[C] \neq 0$ if and only if:

$$\det[A] = 0. \tag{5.44}$$

From Eq. (5.44) values of critical speeds can be calculated. However, those values cannot be obtained in a closed analytical form and have to be calculated numerically. The characteristic equation is solved numerically for a range of the non-dimensional stiffness and three values of slenderness and values of first four critical speeds are plotted in Fig. 5.8. Again, circular rotor cross-section ($\chi = 10/9$) and the Poisson's ratio of 0.3 were assumed.

When the bearings' stiffness is small compared to the rotor stiffness (the soft-mounted case) the stiffness of the bearings has influence only on values of the low-order, rigid critical speeds. Values of the higher order critical speeds, which certainly include flexural modes, are primarily influenced by the shaft slenderness.

In the context of hard-mounted rotors, the concept of 'rigid critical speeds' becomes irrelevant. Flexural vibrations occur together with other, translational and conical movements of the rotor at all critical speeds. When $k^* \to$ inf, value of the first critical speed of the Timoshenko beam can be determined in a closed form according to Eq. (5.36).

The rotor slenderness remains the most critical factor for stable rotation of the rotor at the maximum desired speed. Hence, for a given maximum working speed of

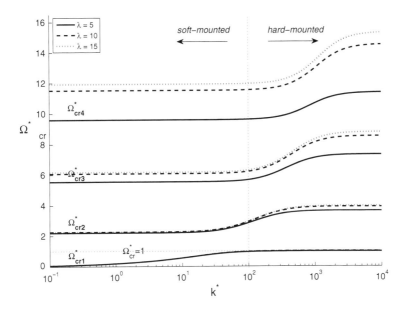

Fig. 5.8 First four critical speeds of a shaft vs bearing stiffness for different values of slenderness ©2010 IEEE

a machine, maximum slenderness λ_{max} of the rotor can be evaluated so that the rotor operates at speeds below the first flexural critical speed (as pointed out in Sect. 2.5).

In today's practice, critical speeds of rotor-bearings systems are usually determined using finite element modeling (FEM). However, in the case of relatively slender rotors, results from this analytical modeling are in good agreement with FEM and measurements, as can be seen in [19, 20].

Finally, the gyroscopic moment resulting from additional mass on the rotor (such as a rotor disc or press fitted elements) also influences values of critical speeds. In the example of asymmetrical rotors, whose center of gravity is not collocated with the center of the shaft, the gyroscopic moment causes an increase of critical speeds, the effect known as *gyroscopic stiffening* (see the Sect. 5.5). However, this effect has importance only with short and thick, gyroscopic rotors for which the stability threshold at the first flexural critical speed is too high to be reached and it ceases to be the limiting factor for the rotational speed.

5.5 Rigid-Rotor Dynamics

In the previous sections dynamical stability limit and critical speeds of, generally compliant, rotors was studied. Although, mathematically, no apparent distinction between rigid and flexural vibration modes was made, based on qualitative analysis in

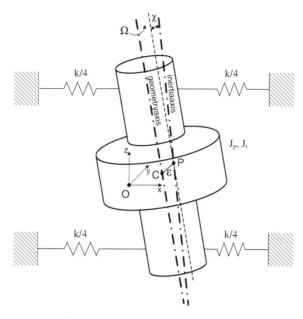

Fig. 5.9 Rigid rotor in compliant bearings

Sect. 5.3 practical dynamical limit of electrical machines was connected with the first flexural critical speed. Since for soft-mounted rotors this critical speed is influenced solely by the rotor slenderness, it can be correlated with the flexural critical speeds and viewed as the limiting machine parameter when it comes to rotordynamical stability.

In this section, dynamic behavior of rigid, thus, infinitely stiff rotors will be analyzed with particular attention to two important phenomena: (rigid) critical speeds and unbalance response. The Jeffcott rotor model will be extended in this section to include rotor moments of inertia and, in turn, gyroscopic effect. Additionally, in order to analyze unbalance response, *static* and *couple unbalance* will be introduced into the rotor model.

In Fig. 5.9 a rigid rotor in compliant bearings is presented. Total radial stiffness of the bearings is k. The rotor spins with an angular speed in axial direction: $\Omega_z = \Omega$. The rotor is characterized with its mass m, polar and transversal moment of inertia J_p and J_t, respectively, and its length and position of the center of mass. It is assumed that geometrical axis of the rotor does not coincide with its inertia axis. The distance ϵ between center of geometry C and center of mass P represents rotor static unbalance while the angle χ represents the couple unbalance—inclination of the inertia axis with respect to geometrical axis.

Figure 5.10 shows arrangement of relevant reference frames and variables which will be used in the modeling. The frame xyz is the stationary reference frame and $\xi\eta\zeta$ is the frame connected to the rotor with its geometry axis coinciding with the ζ-axis. The static unbalance is assumed to lead the couple unbalance for the angle α

Fig. 5.10 Reference frames
used in the modeling

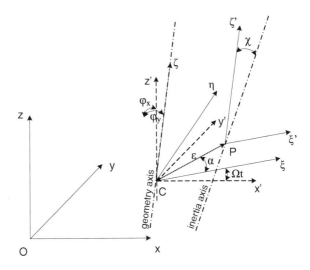

(the angle χ resides in a plane parallel to the $\xi\zeta$ plane). Movements of the rotor will be represented with respect to the stationary frame, namely, using displacements of the rotor's center C: $x_C = x$, $y_C = y$, $z_C = z$, and inclinations of the geometry axis: φ_x, φ_y.

In order to study dynamical behavior of the rigid rotor, the static and couple unbalances will be represented as external forces which spin with the same angular speed as the rotor and which act on an ideal, thus, perfectly-balanced rotor. Equations of movements of an undamped, spinning rotor in the stationary reference frame are expressed as[4] [1]:

$$
\begin{aligned}
m\ddot{x} + k_{11}x + k_{12}\varphi_y &= m\varepsilon\Omega^2 \cos\left(\Omega t + \alpha\right), \\
m\ddot{y} + k_{11}y - k_{12}\varphi_x &= m\varepsilon\Omega^2 \sin\left(\Omega t + \alpha\right), \\
J_t\ddot{\varphi}_x + J_p\Omega\dot{\varphi}_y - k_{12}y + k_{22}\varphi_x &= -\chi\Omega^2\left(J_t - J_p\right)\sin\left(\Omega t\right), \\
J_t\ddot{\varphi}_y - J_p\Omega\dot{\varphi}_x + k_{12}x + k_{22}\varphi_y &= \chi\Omega^2\left(J_t - J_p\right)\cos\left(\Omega t\right).
\end{aligned}
\tag{5.45}
$$

In Eq. (5.45) the terms which include elements of the stiffness matrix k_{11}, $k_{12} = k_{21}$ and k_{22} represent forces that are linked to elastic reaction of the bearings. For the example of the rigid rotor from Fig. 5.9 these stiffness terms can be calculated by inverting the compliance matrix B:

$$
K = \begin{bmatrix} k_{11} & k_{12} \\ k_{12} & k_{22} \end{bmatrix} = B^{-1} = \begin{bmatrix} b_{11} & b_{12} \\ b_{12} & b_{22} \end{bmatrix}^{-1}.
\tag{5.46}
$$

---

[4] Equations of this form can also represent rotors with a massless compliant shaft supported by (generally) compliant bearings.

Fig. 5.11 Position of the center of mass in a cross-section of the rotor

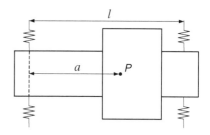

Elements of the compliance matrix can be found using adequate theoretical (or practical) experiments: b_{11} is the displacement of the center of mass P caused by a unit force applied at the same point, b_{12} is the rotation caused by the same force (or, alternatively, displacement cause by a unit moment) and b_{22} is the rotation of the rotor in the center of mass caused by a unit moment at the same point. For the example from Fig. 5.9 the compliance matrix is as follows:

$$B = \begin{bmatrix} \frac{2}{k}\left(1 - 2\frac{a}{l} + 2\left(\frac{a}{l}\right)^2\right) & \frac{2}{kl}\left(1 - 2\frac{a}{l}\right) \\ \frac{2}{kl}\left(1 - 2\frac{a}{l}\right) & \frac{4}{kl^2} \end{bmatrix} \tag{5.47}$$

and the stiffness matrix is calculated accordingly. In (5.47) l represents axial distance between the bearings and a defines the position of the center of mass (refer to Fig. 5.11).

After introducing complex rotor displacement and inclination:

$$r = x + iy,$$
$$\varphi = \varphi_y - i\varphi_x, \tag{5.48}$$

the Eqs. (5.45) can be presented in a matrix form:

$$\underbrace{\begin{bmatrix} m & 0 \\ 0 & J_t \end{bmatrix}}_{\underline{M}} \begin{bmatrix} \ddot{r} \\ \ddot{\varphi} \end{bmatrix} - i\Omega \underbrace{\begin{bmatrix} 0 & 0 \\ 0 & J_p \end{bmatrix}}_{\underline{G}} \begin{bmatrix} \dot{r} \\ \dot{\varphi} \end{bmatrix} + \underbrace{\begin{bmatrix} k_{11} & k_{12} \\ k_{12} & k_{22} \end{bmatrix}}_{\underline{K}} \underbrace{\begin{bmatrix} r \\ \varphi \end{bmatrix}}_{\underline{q}} = \Omega^2 \underbrace{\begin{bmatrix} m\varepsilon e^{i\alpha} \\ \chi\left(J_t - J_p\right) \end{bmatrix}}_{\underline{f}} e^{i\Omega t}$$

$$\tag{5.49}$$

The damping can also be included into the matrix Eq. (5.49) in the following way [1, 21]:

$$\underline{M}\ddot{\underline{q}} + \left(\underline{C}_n + \underline{C}_r - i\Omega\underline{G}\right)\dot{\underline{q}} + \left(\underline{K} - i\Omega\underline{C}_r\right)\underline{q} = \Omega^2 \underline{f} e^{i\Omega t} \tag{5.50}$$

where matrices of non-rotating and rotating damping $\underline{C}_n$ and $\underline{C}_r$, respectively, have a form similar to the form of the stiffness matrix [1]. Introducing the damping into the problem, however, would bring about a great complexity into this study [4, 21].

Moreover, damping barely influences positions of the critical speeds (as seen also in Sect. 5.3) while qualitative insight into the unbalance response can be well drawn using the adopted model of the undamped rotor.

5.5.1 Rigid Critical Speeds

In order to find critical speeds of the undamped system from Fig. 5.9, the case of rotor free whirling will be studied. One should, therefore, look for homogeneous solution of the differential Eq. (5.49), thus:

$$\underbrace{\begin{bmatrix} m & 0 \\ 0 & J_t \end{bmatrix}}_{M} \begin{bmatrix} \ddot{r} \\ \ddot{\varphi} \end{bmatrix} - i\Omega \underbrace{\begin{bmatrix} 0 & 0 \\ 0 & J_p \end{bmatrix}}_{G} \begin{bmatrix} \dot{r} \\ \dot{\varphi} \end{bmatrix} + \underbrace{\begin{bmatrix} k_{11} & k_{12} \\ k_{12} & k_{22} \end{bmatrix}}_{K} \underbrace{\begin{bmatrix} r \\ \varphi \end{bmatrix}}_{q} = 0 \qquad (5.51)$$

After introducing a solution of type:

$$\underline{q} = \underline{q_0} e^{st} = \begin{bmatrix} r_0 \\ \varphi_0 \end{bmatrix} e^{st} \qquad (5.52)$$

the following system of equations is obtained:

$$\begin{aligned} \left(ms^2 + k_{11} \right) r_0 + k_{12} \varphi_0 &= 0, \\ k_{12} r_0 + \left(J_t s^2 - i s J_p \Omega + k_{22} \right) \varphi_0 &= 0. \end{aligned} \qquad (5.53)$$

The system (5.53) has non-trivial solutions if and only if:

$$\det \begin{bmatrix} ms^2 + k_{11} & k_{12} \\ k_{12} & J_t s^2 - i s J_p \Omega + k_{22} \end{bmatrix} = 0. \qquad (5.54)$$

Since there is no damping in the system, the solutions of the Eq. (5.54) are purely imaginary. After substitution of $s = i\omega$, the Eq. (5.54) becomes:

$$\omega^4 m J_t - \omega^3 \Omega m J_p - \omega^2 \left(k_{11} J_t + m k_{22} \right) + \omega \Omega k_{11} J_p + k_{11} k_{22} - k_{12}^2 = 0, \quad (5.55)$$

and it gives an implicit correlation between the natural frequencies of the system and the rotational speed based on which a Campbell diagram can be plotted.

To find values of critical speeds, $\omega = \Omega = \Omega_{cr}$ is substituted into (5.55) and that leads to the following biquadratic equation over the critical speed:

$$m \left(J_p - J_t \right) \Omega_{cr}^4 - \left[k_{11} \left(J_p - J_t \right) - m k_{22} \right] \Omega_{cr}^2 - \left(k_{11} k_{22} - k_{12}^2 \right) = 0 \qquad (5.56)$$

Positive solutions of Eq. (5.56) are given by:

$$\Omega_{cr} = \sqrt{\frac{k_{11}(J_p - J_t) - mk_{22} \pm \sqrt{[k_{11}(J_p - J_t) - mk_{22}]^2 + 4(k_{11}k_{22} - k_{12}^2)m(J_p - J_t)}}{2m(J_p - J_t)}}$$

(5.57)

It can be noticed that two real solutions for Ω_{cr} exist only if transversal moment of inertia is higher than the polar moment ($J_t > J_p$); in the opposite case ($J_p > J_t$) only one real solution exist. Rotors whose polar moment of inertia is higher than the transversal are referred to as *short* or *disc-shaped* rotors and they have one critical speed less than *long* (slender) rotors.

Using (5.46) and (5.47) it can be shown that the stiffness matrix of the rigid rotor from Fig. 5.9 can be expressed in the following form:

$$K = k \begin{bmatrix} k'_{11} & lk'_{12} \\ lk'_{12} & l^2 k'_{22} \end{bmatrix},$$

(5.58)

where terms k'_{11}, k'_{12} and k'_{22} depend solely on the ratio a/l (Fig. 5.11).

After replacing the stiffness terms of (5.57) with the adequate elements of the stiffness matrix (5.58) values of critical speeds can be expressed in the following non-dimensional form:

$$\Omega^*_{cr} = \sqrt{\frac{\Gamma^* k'_{11} - k'_{22} \pm \sqrt{\left(\Gamma^* k'_{11} - k'_{22}\right)^2 + 4\left(k'_{11} k'_{22} - k'^2_{12}\right)\Gamma^*}}{2\Gamma^*}},$$

(5.59)

where Ω^*_{cr} is ratio between the critical rotational speed and the critical speed of the undamped Jeffcott rotor:

$$\Omega^*_{cr} = \frac{\Omega_{cr}}{\sqrt{k/m}},$$

(5.60)

and Γ^* is a non-dimensional equivalent of the gyroscopic moment [1]:

$$\Gamma^* = \frac{(J_p - J_t)}{ml^2}.$$

(5.61)

The critical speeds calculated from Eq. (5.59) are plotted in Fig. 5.12. Again, it is evident that short rotors are associated with only one critical speed. Cylindrical and conical movement of rotors are generally coupled (unless $a/l = 0.5$) and so are the corresponding vibrations. Still, vibrations of the rotor are usually dominantly cylindrical or conical and the lower critical speed is usually considered cylindrical while the higher one (if existent) is considered conical.

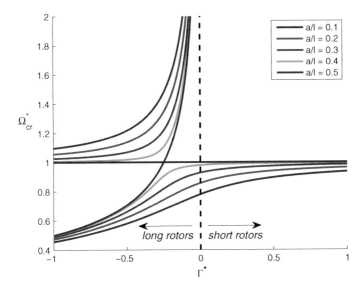

Fig. 5.12 Critical speeds of rigid rotors with respect to the gyroscopic moment

In the special case of a symmetrical rotor whose center of mass is in the middle of the shaft ($a/l = 0.5$) critical speeds take on the following values:

$$\Omega_{cr,cyl} = \sqrt{\frac{k_{11}}{m}} = \sqrt{\frac{k}{m}}, \tag{5.62}$$

$$\Omega_{cr,con} = \sqrt{\frac{k_{22}}{(J_t - J_p)}} = \frac{l}{2}\sqrt{\frac{k}{(J_t - J_p)}}. \tag{5.63}$$

Naturally, conical critical speed is imaginary in the case of short rotors.

Gyroscopic effect influences rise of critical speeds; the phenomenon which is sometimes referred to as "gyroscopic stiffening" although no real stiffening is taking place [5, 22]. It should be pointed out that the non-dimensional gyroscopic moment of most of realistic rigid rotors lie with in a small range around zero. Namely, if the gyroscopic moment is very small ($J_t \gg J_p$), the rotor is too slender to be considered rigid; conversely, if $J_p \gg J_t$ the rotor is practically a thin disc and again cannot be regarded as rigid [1].

Finally, a special case that should be considered is the case of rotors with $J_t \approx J_p$. This rotor type is highly unsuitable for high speed rotation. Namely, if a Campbell diagram is plotted based on Eq. (5.55) it can be shown [1] that, at high speeds, the resonant frequency associated with the forward conical mode asymptotically approaches the line $\omega = \Omega J_p/J_t$. If $J_p = J_t$ the rotor would always be in the vicinity of the critical speed (although never *at* the critical speed) and the vibrations would steadily grow as the rotor speed increases. The rotor would not have a preferred position and would not be able to benefit from *self-centering* at high speeds (see the Sect. 5.5.2).

5.5.2 Unbalance Response

Dynamic response of the rotor to the unbalance is found as the particular solution of the differential Eq. (5.49). The equation is presented here again in a compact form:

$$M\underline{\ddot{q}} - i\Omega G\underline{\dot{q}} + K\underline{q} = \Omega^2 \begin{bmatrix} m\varepsilon e^{i\alpha} \\ \chi\left(J_t - J_p\right) \end{bmatrix} e^{i\Omega t} \tag{5.64}$$

The solution can be found in the following form:

$$\underline{q} = \underline{q}_0 e^{i\Omega t} = \begin{bmatrix} r_0 \\ \varphi_0 \end{bmatrix} e^{i\Omega t}, \tag{5.65}$$

which, after substitution into (5.64), leads to the expressions for amplitudes of rotor's cylindrical and conical whirl:

$$\underline{q}_0 = \begin{bmatrix} r_0 \\ \varphi_0 \end{bmatrix} = \Omega^2 \left[-\Omega^2\left(\underline{M} - \underline{G}\right) + \underline{K}\right]^{-1} \begin{bmatrix} m\varepsilon e^{i\alpha} \\ \chi\left(J_t - J_p\right) \end{bmatrix}. \tag{5.66}$$

It can be inferred from (5.66) that the amplitudes r_0 and φ_0 are, in general, complex numbers since the static and couple unbalance do not lie in the same plane. Yet, in other to make important conclusions about the unbalance response, it is sufficient to study the case in which phase angle of the static unbalance vanishes ($\alpha = 0$). In that case, the last equations yields the following expression:

$$\underline{q}_0 = \frac{\Omega^2}{\Delta} \begin{bmatrix} m\varepsilon\left(k_{22} - \left(J_t - J_p\right)\Omega^2\right) - \chi\left(J_t - J_p\right)k_{12} \\ -m\varepsilon k_{12} + \chi\left(J_t - J_p\right)\left(k_{11} - m\Omega^2\right) \end{bmatrix}, \tag{5.67}$$

where:

$$\Delta = m\left(J_t - J_p\right)\Omega^4 - \left(k_{11}\left(J_t - J_p\right) + mk_{22}\right)\Omega^2 + k_{11}k_{22} - k_{12}^2. \tag{5.68}$$

If the elements of the stiffness matrix are represented according to (5.58), considering (5.61), the amplitudes of the unbalance response can be expressed using non-dimensional quantities:

$$\begin{bmatrix} \dfrac{r_0}{\varepsilon}\bigg|_{\chi=0} & \dfrac{r_0}{\chi l}\bigg|_{\varepsilon=0} \\ \dfrac{\varphi_0}{\varepsilon/l}\bigg|_{\chi=0} & \dfrac{\varphi_0}{\chi}\bigg|_{\varepsilon=0} \end{bmatrix} = \frac{\Omega^{*2}}{\Delta^*} \begin{bmatrix} k'_{22} + \Gamma^*\Omega^{*2} & \Gamma^* k'_{12} \\ -k'_{12} & -\Gamma^*\left(k'_{11} - \Omega^{*2}\right) \end{bmatrix}, \tag{5.69}$$

where the rotational speed is rated by the critical speed of the Jeffcott rotor:

$$\Omega^* = \frac{\Omega}{\sqrt{k/m}}, \tag{5.70}$$

and non-dimensional determinant is defined as:

$$\Delta^* = \frac{\Delta}{k^2 l^2} = -\Gamma^* \Omega^{*4} + (\Gamma^* k'_{11} - k'_{22})\Omega^{*2} + k'_{11} k'_{22} - k^2_{12}. \tag{5.71}$$

For high-speed rotors it is particularly important to analyze the rotor behavior in the supercritical regime. As the speed increases, the influence of cross-coupling terms of Eq. (5.69) becomes immaterial. Here, a special case of the symmetrical rotor will be examined whose center of mass is in the middle of the shaft or, equivalently, whose cylindrical and conical movements are decoupled ($k_{12} = 0$). The rotor of the test machine belongs to this category.

From (5.46), (5.47) and (5.58) for the rigid symmetrical rotor it holds $k'_{11} = 1$ and $k'_{22} = 1/4$. Hence, expressions for amplitude response to static and couple unbalance reduce to:

$$\frac{r_0}{\varepsilon} = \frac{\Omega^{*2}}{1 - \Omega^{*2}}, \tag{5.72}$$

$$\frac{\varphi_0}{\chi} = -\frac{\Omega^{*2}}{\Omega^{*2} + \frac{1}{4\Gamma^*}}. \tag{5.73}$$

Amplitudes of the unbalance response are plotted against rotational speed in Figs. 5.13 and 5.14 according to expressions (5.72) and (5.73). In supercritical regions amplitudes of the rotor's translational and conical movements tend to $-\varepsilon$ and $-\chi$ respectively. Physically, it means that the rotor changes its axis of rotation from geometrical to inertia axis as the speed advances in the supercritical regime. This phenomenon is referred to as *self-centering* and it is a result of the fact that in the supercritical region inertia forces overcome bearing forces causing the rotor to spin around its principal axis of inertia.

In the example of the response to static unbalance and also to couple unbalance in long rotors the change of preferred axis of rotation happens very fast as the rotor crosses the critical speed: the whirl of the geometric center in the stationary reference frame inverts its phase with respect to the excitation. In the example of short rotors, on the other hand, the alignment of the rotor to couple unbalance proceeds gradually.

Self-centering is a greatly exploited phenomenon in design of supercritical rotating machinery [1, 20, 22]. Instead of making a geometrically perfect rotor in a fixed support, it is often more reasonable to arrange a flexible support for the rotor and let the rotor inertia offset the influence of the unbalance at high speeds.

Distinguishing case is, however, rotors with $\Gamma^* = 0$ ($J_p = J_t$, see Fig. 5.14). Such a rotor type cannot benefit from self-alignment and is generally avoided when it comes to supercritical rotors.

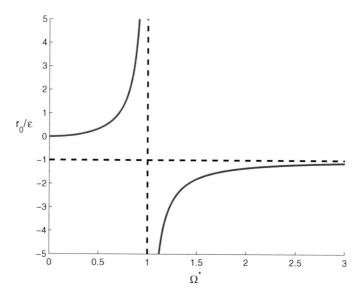

Fig. 5.13 Amplitude of response to static unbalance of symmetrical rotors

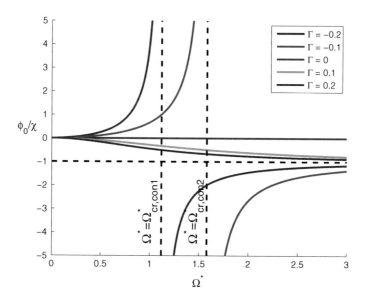

Fig. 5.14 Amplitude of response to couple unbalance of symmetrical rotors

It is important to mention that the model of an undamped rotor is not adequate to describe rotor behavior in the vicinity of a critical speed. Unbalance response at a critical speed highly depends on damping which attenuates the vibrations and the response is also influenced by the rate of the speed increase through the critical speed [5].

5.6 Conclusions

This chapter studies vibrations of a rotor of a high-speed electrical machine, their cause and influence on the rotor-bearings system. Two types of vibrations are distinguished: resonant (forced) and self-excited; the latter vibrations draw particular concern since they are unstable and hazardous. The speed at which self-excited vibration occur is referred to as instability threshold speed and it represents rotordynamical limit for the rotor rotational speed.

Rotors of high-speed machines are usually soft-mounted meaning that their stiffness is much higher than the stiffness of their bearings. For such rotors it is possible to recognize two modes of vibrations based on whether the rotor is deformed while vibrating: rigid-body and flexural (bending) vibration modes. Using a simple, Jeffcott rotor model, the chapter shows that rotation can become unstable in the supercritical regime of a certain vibration mode if rotating damping of the rotor-bearing system affects that mode.

Rotors are usually stable in a supercritical speed range which corresponds to rigid-body vibrational modes and today's high-speed rotors regularly operate in that speed range. An exception are rotors supported by bearings with a non-synchronous damping such as lubricated journal bearings and hydrodynamic bearings.

Rotors which possess some internal damping can easily become unstable in a supercritical range corresponding to flexural modes. Rotors of electrical machines are receptive to eddy-currents, always comprised of fitted elements and often contain materials, such as composites, with significant material damping. Therefore, these rotors are prone to be unstable in flexural supercritical regimes. The first flexural critical speed practically represents the rotordynamical speed limit of an electrical machine.

Critical speeds of a high-speed rotor (represented as a cylindrical Timoshenko beam) are correlated with the rotor slenderness and bearing stiffness. It was shown that flexural critical speeds of a rotor-bearings system depend solely on the rotor slenderness. The results of the analytical modeling comply with FEM calculations available in literature [19]. Flexural critical speeds are far above the operating speed range of the test motor and the calculations of these speeds are irrelevant for the test machine.

The rotor slenderness is the most critical factor for stable rotation of a rotor in a high-speed electrical machine. Hence, a maximum slenderness of the rotor can be evaluated so that the rotor operates at speeds below the first flexural critical speed.

Last section of the chapter analyzes behavior of rigid rotors i.e. rotors which operate well below the speeds in which flexural vibrations occur. For the analysis purpose, the Jeffcott rotor model is extended to include gyroscopic effect.

Rotors whose polar moment of inertia is higher than the transversal are referred to as short or disc-shaped rotors and they have one critical speed less than long (slender) rotors. Gyroscopic effect influences rise of critical speeds; the phenomenon which is sometimes referred to as *gyroscopic stiffening*.

At supercritical regime, rigid rotors change their axis of rotation from geometrical to inertia axis as the speed advances. This phenomenon is referred to as *self-centering* and it can be exploited in design of supercritical rotating machinery: instead of making a geometrically perfect rotor in a fixed support, it is often more reasonable to arrange a flexible support for the rotor and let the rotor inertia offset the influence of the unbalance at high speeds.

The dynamic of a rotor-bearings system and its important aspects—stability of rotation and critical speeds—presented in this chapter are fairly known and well-researched in the field of rotordynamics; most models can be found in textbooks. The importance and main contribution of this chapter is that it highlights the effects of those phenomena relevant to rotational stability of electrical machines.

References

1. G. Genta, *Dynamics of Rotating Systems* (Springer, Berlin, 2005)
2. F. Ehrich, D.W. Childs, Self-excited vibrations in high performance turbomachinery. Mech. Eng. **106**(5), 66–79 (1984)
3. G. Genta, E. Brusa, On the role of nonsynchronous rotating damping in rotordynamics. Int. J. Rotat. Mach. **6**(6), 467–475 (2000)
4. G. Genta, F. De Bona, Unbalance response of rotors: a modal approach with some extensions to damped natural systems. J. Sound Vib. **140**(1), 129–153 (1990)
5. D. Childs, D.W. Childs, *Turbomachinery Rotordynamics: Phenomena, Modeling, and Analysis* (Wiley-Interscience, New York, 1993)
6. A. Muszynska, Whirl and whip-rotor/bearing stability problems. J. Sound Vib. **110**(3), 443–462 (1986)
7. A. Muszynska, Stability of whirl and whip in rotor/bearing systems. J. Sound Vib. **127**(1), 49–64 (1988)
8. A. Borisavljevic, H. Polinder, J. Ferreira, On the speed limits of permanent-magnet machines. IEEE Trans. Ind. Electron. **57**(1), 220–227 (2010)
9. V. Kluyskens, B. Dehez, H. Ahmed, Dynamical electromechanical model for magnetic bearings. IEEE Trans. Magn. **43**(7), 3287–3292 (2007)
10. J. Melanson, J.W. Zu, Free vibration and stability analysis of internally damped rotating shafts with general boundary conditions. J. Vib. Acoust. **120**, 776 (1998)
11. W. Kim, A. Argento, R. Scott, Forced vibration and dynamic stability of a rotating tapered composite timoshenko shaft: Bending motions in end-milling operations. J. Sound Vib. **246**(4), 583–600 (2001)
12. T. Kenull, W.R. Canders, G. Kosyna, Formation of self-excited vibrations in wet rotor motors and their influence on the motor current. Eur. Trans. Electr. Power **13**(2), 119–125 (2007)
13. S.H. Crandall, A heuristic explanation of journal bearing instability, in *Proceedings of the Workshop on Rotordynamic Instability Problems in High-Performance Turbomachinery, Texas A&M University, College Station, Texas*, pp. 274–283, 1982
14. T. Iwatsubo, Stability problems of rotor systems. Shock Vib. Inform. Digest (Shock and Vibration Information Center) **12**(7), 22–24 (1980)
15. B. Murphy, S. Manifold, J. Kitzmiller, Compulsator rotordynamics and suspension design. IEEE Trans. Magn. **33**(1), 474–479 (1997)
16. B.H. Rho, K.W. Kim, A study of the dynamic characteristics of synchronously controlled hydrodynamic journal bearings. Tribol. Int. **35**(5), 339–345 (2002)
17. E. Brusa, G. Zolfini, Non-synchronous rotating damping effects in gyroscopic rotating systems. J. Sound Vib. **281**(3–5), 815–834 (2005)
18. D. Guo, F. Chu, D. Chen, The unbalanced magnetic pull and its effects on vibration in a three-phase generator with eccentric rotor. J. Sound Vib. **254**(2), 297–312 (2002)

19. J. Ede, Z. Zhu, D. Howe, Rotor resonances of high-speed permanent-magnet brushless machines. IEEE Trans. Ind. Appl. **38**(6), 1542–1548 (2002)
20. T. Wang, F. Wang, H. Bai, H. Cui, Stiffness and critical speed calculation of magnetic bearing-rotor system based on fea, in *Electrical Machines and Systems, 2008. ICEMS 2008. International Conference on*, pp. 575–578, 17–20 Sept 2008
21. J.T. Sawicki, G. Genta, Modal uncoupling of damped gyroscopic systems. J. Sound Vib. **244**(3), 431–451 (2001)
22. M. Kimman, H. Langen, J. van Eijk, H. Polinder, Design of a novel miniature spindle concept with active magnetic bearings using the gyroscopic stiffening effect, in *Proceedings of the 10th International Symposium on Magnetic Bearing, Martigny, Switzerland*, 2006

Chapter 6
Bearings for High-Speed Machines

6.1 Introduction

Advance in electrical machines is characterized, among other, by pursuit of ever higher rotational speeds, particularly regarding turbomachinery and machining spindles [1]. Demands for increasing tangential speeds and, often, positioning accuracy could not be achieved by using standard journal or ball bearings. Therefore, the requirement for speed of the machinery brought about improvements in bearing technology. Enhancements have been either sought within the standard (mechanical) bearing technology that would become suitable for required high speeds or alternatives have been looked for in the form of contactless bearings.

Speed capability of rotational bearings is usually represented in terms of the DN number which represents product of the bearing inner diameter in mm and rotational speed of the rotor in rpm. Standard ball bearings have the DN number below 500.000 [2]. On the other hand, DN values that are currently needed in certain high-tech applications such as turbines for aircraft or high-speed spindles amount to a few millions [3]. Under such conditions mechanical bearings are subjected to great centrifugal loading and high temperatures which both reduce bearings load capacity and life-time [3].

Bearing type and stiffness have strong influence on rotational accuracy which is of particular importance in machining spindles. The accuracy can be improved, though, by performing rotor balancing, the process that may be time-consuming and costly, or by exploiting *self-centering* of high-speed rotors if low-stiffness bearings are used (see the Chap. 5).

Capability of bearings to support required speed and accuracy of the machine along with their cost and durability play the most important role in choice of bearings for high-speed machines.

Goal of this chapter is to study different types of bearings with respect to their applicability for high-speed rotation. The chapter is primarily concerned with the bearings that have been the most promising for high speed: (hybrid) ball bearings, externally pressurized (or static) air bearings and magnetic bearings. A general overview and comparisons will be given in the end of the chapter.

A. Borisavljević, *Limits, Modeling and Design of High-Speed Permanent Magnet Machines*, 117
Springer Theses, DOI: 10.1007/978-3-642-33457-3_6,
© Springer-Verlag Berlin Heidelberg 2013

6.2 Mechanical Bearings

Conventional, mechanical bearings are still predominantly used among commercial machines. The greatest advantages of mechanical bearings are robustness and low cost. However, they have limited operational temperature making it the main restriction for the rotational speed. Furthermore, great increase in speed tremendously intensifies wear due to friction which, in turn, shortens life-time of the bearings.

Particular problem with high-speed rotation is centrifugal loading which rises with square of the speed and lowers load capacity of the bearings [3]. In order to mitigate this problem, rotor balancing is often required. For rigid rotors, though, balancing in two planes to compensate for static and couple unbalance is usually sufficient [4].

Although a lot of effort has been invested, particularly in academia, in developing and promoting contactless (or frictionless) bearings, mechanical bearings are still widely used for high-speed applications due to their simplicity and low cost. For majority of commercial applications conventional bearings will suffice making fluid (air) and magnetic bearings reserved for special applications with stringent requirements [5]. At the same time, considerable research is aimed at improving quality of mechanical bearings [6].

For high-speed applications ball-bearings are mostly used; lubricated journal bearings are generally avoided because of their issues with instability [7, 8]. Today, specially-designed ball bearings can be found for rotational speeds up to 100.000 rpm and values of the DN number as high as 1.5 million [3].

A distinctive example of a very high-speed commercial machine running on a ball bearings is Dyson's 104.000 rpm DC brushless motor (DDM V2 [9]) which has a simple and small, balanced permanent magnet rotor without retainment.

Ball bearings are commonly found in dental spindles where they can support speeds up to 500.000 rpm [10]. When it comes to machining spindles, air and magnetic bearings replace mechanical bearings in precision machinery to an ever greater extent.

Zwyssig et al. from ETH Zurich have published successful designs of several high-speed motors that run on ball bearings up to speed of 1 million rpm [10]. The DN value of the used bearings is around 1.6 million [11], however, no data on durability of those bearings is available.

In high-speed turbomachinery a trend of transition from mechanical to air bearings is noticeable [1]. Ball bearings limited the speed of a gas turbine designed at University of Leuven to 160.000 rpm [12]. The authors suggested that special ball bearings would allow speeds over 200.000 rpm; however, air bearings would be needed for the desired speed of 420.000 rpm (Fig. 6.1).

Improvements of ball bearings consist in reduction of size of the balls and transition to ceramic materials to accommodate high speeds [6]. These changes have been followed by novel designs of geometry, curvature and seals [6]. Oil lubrication is replaced with a small amount of grease and inclusion of the seals allow that bearings are sealed and greased for life [6, 13]. Combination of ceramic balls and grease lubrication allow very long operational time of the bearings with the DN index well

Fig. 6.1 SKF ceramic hybrid bearings

above 1.5 million [14]. Such bearings can also be used at cryogenic temperatures at which standard lubricants would solidify [15].

Exceptionally good result have been obtained with silicon-nitride hybrid bearings which represent bearings consisting of ceramic, silicon-nitride balls and steel rings. Very small wear in these bearings is result of the fact that "silicon nitride rubbing on M-50 steel offers friction and wear characteristics as good as those of silicon nitride rubbing on itself" [16], which results in a very long bearing life. Due to the comparably low density of ceramics, silicon-nitride balls reduce centrifugal loading on the outer, steel, raceway of the bearings [3]. Ceramics, in general, are also less sensitive to lubricant type and lubricant contamination [3].

Authors of [3] report hybrid silicon nitride bearings with DN values up to 4 millions. According to [17] machining spindles can be supported with high stiffness and rotational accuracy by using ceramic hybrid bearings. "In general, compared to all-steel bearings, rolling contact ceramic bearings can more easily meet requirements of higher efficiency, higher speed, higher reliability, higher accuracy, greater stiffness, longer life, marginal lubrication, lower friction, corrosion resistance and non-conductivity and with less maintenance action." [3]

6.3 Air (Fluid) Bearings

Air bearings use a thin fluid film or pressurized air to support the rotor. Fluid (air) forms a layer between the bearing housing and the shaft ("gas lubrication") transferring, at the same time, the force which supports the shaft. This principle of operation has been established for more than 50 years; nowadays, technology of air bearings is quite mature making them frequently applied in a large number of applications.

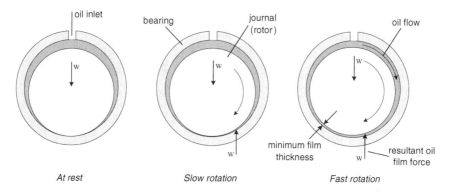

Fig. 6.2 Hydrodynamic bearing

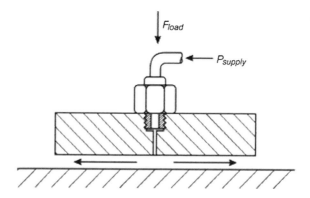

Fig. 6.3 Aerostatic bearing

Because of very small air-gaps (in order of tens of μm) air bearings require flawless geometry with very low tolerances. Extremely high motion accuracy can be, therefore, achieved with high precision and repeatability [18] which makes air bearings very attractive for precision machinery (Figs. 6.2 and 6.3).

According to how the pressure in fluid is generated, two types of air bearings are recognized: static and dynamic. Dynamic bearings use relative motion between the moving body (shaft) and bearing housing to generate hydrodynamic pressure in the fluid. Liquid (oil) is used as the fluid rather then gas (air) because of higher viscosity; these bearings are usually referred to as *hydrodynamic*. Hydrodynamic bearings are essentially frictionless only at high speeds—at zero and low speeds the shaft is in contact with the housing (or rather, with the foil journal lining in foil bearings).

Static air bearings use externally pressurized air to levitate the body (rotor) at all possible speeds including zero speed. From the pressure supply the air is directed through either small holes in the bearing (orifices) or porous material.

For high speed machinery static air bearings are often used. Although having higher load capacity, hydrodynamic bearings can operate efficiently only in a narrow speed range [19] and have larger thermal deformation than aerostatic bearings [18]. Dynamic bearings are preferably used in heavy-load precision machines while static air bearings are used in small and medium, high-speed and/or precision machines [18].

Very high speeds have been achieved in laboratory environments using air bearings in the example of gas turbines. A micromachined air turbine supported with air bearings was designed at MIT and operated at 1.3 million rpm with tangential speed of 300 m/s [20].

High-speed machining spindles with aerostatic bearings have μm motion accuracy at speeds up to 200.000 rpm, such as the micromilling spindle designed at Brunel University [19].

Air bearing can have very high static and dynamic stiffness, comparable with ball bearings [21]. On the other hand, static air bearings generally have low load capacity. Another disadvantage of air bearings lies in the need for preloading for certain geometries and directions. While journal bearings are usually self-preloaded, trust bearings need preloading to increase the stiffness and maintain a fixed air-gap.

The most salient problem with air bearings, though, is stability of rotation. Hydrodynamic journal bearings have a self-excited instability, which is commonly called *whirl instability*, and it is correlated with behavior of the fluid film in the bearings. The fluid in the journal bearings moves at an average speed which is close to half of tangential speed of the rotor [8, 22, 23]. This movement is reflected in subsynchronous whirling of the rotor with frequency $\omega = \Omega/2$ which is superimposed to other whirling motions. At speed near twice of the first critical speed the whirl frequency reaches the value of the first resonant frequency of the system. The whirl is then replaced by *oil whip*—a particularly destructive "lateral forward precessional subharmonic vibration of the rotor" [7]. Independently of the further increase of speed oil whip maintains the constant frequency (see the cascade plot of the rotor vibrations in Fig. 6.4).

Partially, phenomenon of whirl instability was explained in Sect. 5.3.2 using the Jeffcott rotor model: the fluid in the bearing provides a subsynchronously rotating damping which rotates with a speed close to $\Omega/2$. However, the behavior is far more complex; that model cannot account for the phenomenon of oil whip. As Muszynska points out [7]: "Researchers and engineers do not always agree upon the physical description of the shaft/bearing or shaft/seal solid/fluid interaction dynamic phenomena. The complexity of these phenomena and the long list of factors affecting them make the picture tremendously obscure."

A heuristic explanation for the whirl instability is given by Crandall in [8] and analytical models and explanations can be found in papers by Muszynska [7, 24].

Aerostatic bearings have, in general, higher dynamic stiffness and lower viscous drag than hydrodynamic bearings, both properties having positive influence on their stability [25]. Nevertheless, this type of bearings is not immune to the whirl instability [25] since aerodynamic forces completely overwhelm aerostatic forces at very high speeds [19].

In order to avoid instability the system is designed with constraint of having a maximum operating speed of the rotor well below the frequency of whirl instability,

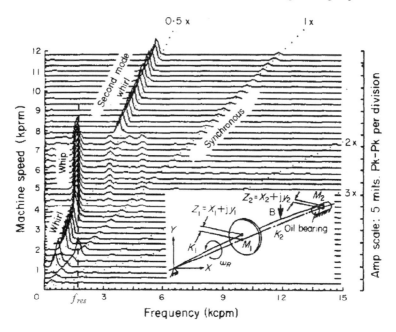

Fig. 6.4 Cascade plot of rotor vibrations measured in an oil-lubricated bearing, taken from [24]: cascade lines represent amplitudes of the rotor whirl at given rotational speeds and frequencies. The plot shows transitions of self-excited vibrations from whirl to whip and again from whip to a second-mode whirl

i.e. double value of the first critical speed [26]. At the same time, bearings can be optimized to maximally increase the threshold of instability [27, 28] and/or innovative bearing designs can be used [29].

Unfortunately, air bearings can suffer from other forms of, mainly pneumatic, instabilities, prevention of which must be considered in the design (see [30, 31]). Emergence of pneumatic instabilities depends on mechanical design of the bearings which is beyond the scope of this chapter.

6.4 Active Magnetic Bearings

Active magnetic bearings (AMB) use force of an electromagnet for levitation of a body (rotor). An active magnetic bearing consists of an electromagnet and a power amplifier that supplies the electromagnet with currents (Fig. 6.5). The bearing needs continuous current input and a closed-loop controller since active magnetic bearings are unstable in open loop. A magnetically levitated body may have certain degrees of freedom (DOF) passively controlled, thus, by means of permanent magnets only; however, for stable levitation at least one DOF must be actively controlled, according to the theorem of Earnshaw [32].

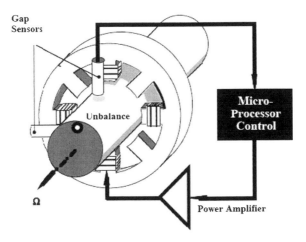

Fig. 6.5 Basic setup of an active magnetic bearing carrying a rotor [33]

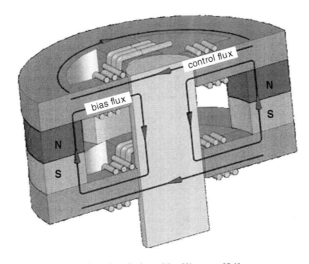

Fig. 6.6 A homopolar magnetic bearing designed by Kimman [34]

As a rule, rotors are supported by pairs of AMB in differential mode that enables linear control of the magnetic force and, in turn, rotor position. Quite often, a permanent magnet is utilized to create the bias flux in the bearings and current of the coils is then used to control the rotor position. The most suitable for support of high speed rotors is the *homopolar* bearing structure in which the rotor (ideally) does not experience changes of the bias flux while rotating (Fig. 6.6).

For a long time magnetic bearings were considered too complex and expensive to be commercially appealing [35]. However, in course of the last two decades AMB have proven their effectiveness and reliability and shown great potential for an increasing number of applications [33, 35]. AMB have several important advantages with respect to other types of bearings. They provide purely contact- and frictionless operation with no contamination and no need for lubricants.

Beside frictionless operation, AMB offer possibility of creating practically arbitrary damping or stiffness, the property which can be greatly utilized to adjust the dynamical properties of the system as the rotational speed changes.

All these properties make AMB particularly attractive for high speed applications. They do not suffer from instabilities connected with air or lubricant flow as fluid bearings do. Their stiffness is generally lower than the stiffness of air bearings; however, having larger clearances, AMB rotors can exploit self-centering and operate stably at supercritical speeds.

Still, magnetic bearings are rather complex electromechanical systems which include sensors, power electronics/amplifiers, advanced digital controllers and electromagnets that, in the end, affect the price and reliability of the bearings. Their stiffness declines with the operating frequency [33] and they have limited control bandwidth [36] due to presence of coils and sensors and limited processing power of controllers. Gyroscopic effect and/or non-linearities can destabilize the rotor-bearings system at high speeds [37] and more complex control methods are needed such as cross-feedback [37] or non-linear feedback control [32].

The benefit of adaptable damping and stiffness has virtually no influence on high-frequency whirl and flexural vibrations. AMB are usually located at points where flexural vibrations are very small. Furthermore, the frequency range in which AMB can counteract forces that cause displacement of the rotor is limited by the control loop bandwidth [38]. Frequencies of flexural vibrations of high-speed rotors are usually far above the bandwidth of the bearings so the vibrations cannot be suppressed.

Nevertheless, AMB technology is advancing and the bearings appear in many precision and high speed applications [39]. Improvements have been made in design of specialized power amplifiers and their integration with AMB [35], control algorithms [40] and design of AMB for high speed [36].

There have been an increasing number of AMB in turbomachinery for which low maintenance and long life-time of AMB under sever conditions is a great advantage [41–43]. In general, turbomachinery is currently the main commercial application of AMB [39] among rotating machines although improving reliability of AMB is a key requirement for making them widely embraced by this industry.

AMB are considered less mature technology for high-speed production machinery in comparison with air bearings; however, high-speed machining spindles are a promising application of magnetic bearings. Considerable research on AMB for machining applications has been conducted at University of Virginia [44] and at TU Delft [36]. Hybrid solutions have also emerged in high-speed machining spindles where AMB have been combined with other types of bearings and actuators to take advantage of certain virtues of AMB such as accurate positioning during rotation [45].

Today, AMB are still mainly used in rather special applications where bearings either need to work under special conditions (vacuum, harsh environment [39]) or must conform to very strict limitations (no lubrication, no contamination [5]). An

Table 6.1 Advantages and disadvantages of different bearing types

Ball bearings	Air bearings	Active magnetic bearings
+ Low cost	+ No friction/wear	+ No friction/wear
+ Robust	+ Ultra precision	+ Zero contamination
	+ High stiffness	+ No maintenance
	+ Low maintenance	+ Adjustable force, damping
		+ Positioning during rotation
		+ Modular design
		+ Operate in harsh settings
		+ Facilitate monitoring
− Temperature limited	− Instabilities	− Complex
− Wear	− Low load capacity	− Expensive
− Need lubrication	− Require flawless geometry	− Low reliability
− Need maintenance	− Need preloading	− Require control
	− Susceptible to dirt, temperature	− Require constant power supply, sensors, electronics

example of such a special application is 60.000 rpm, 4.1 kW magnetically levitated flywheel which NASA intends to use as a replacement for batteries on the space [46].

Finally, AMB find use in many important additional tasks such as force monitoring [47], system identification [48], machine state diagnosis [49], monitoring and suppressing vibration levels [38], etc.

6.5 Conclusions

In this chapter different bearing types have been studied with respect to their applicability in high-speed machinery.

Ball bearings are still a predominant bearing type among commercial high-speed machines due to their robustness and low costs. Different types of ball bearings are regularly present in machines with rotational speeds up to 100.000 rpm. Hybrid ball bearings with silicon-nitride balls represent the most promising type of mechanical bearings in terms of not only speed performance but also reliability, stiffness, lifetime and low contamination.

Nevertheless, ball bearings have apparent limitations for extremely high rotational and tangential speeds and alternatives are sought in different types of contactless bearings.

Air bearings are well-known and widely utilized technology for frictionless support of rotors. Air bearings offer high stiffness and great accuracy and repeatability of rotation which makes them ideal for precision machinery. Aerostatic bearings are particularly suitable for low- and middle volume high-speed machines. However, a salient problem of air bearings are instabilities which limit rotational speed and complicate design of the bearings.

Active magnetic bearings seem to have the greatest potential for high-speed applications: perfect conditions for rotation (no friction, wear, lubricants), modularity in design and possibility of regulating stiffness and damping of the bearings due to their active nature. On the other hand, they are not seldom considered too complex, costly and unreliable. Nowadays, they are mostly used in special applications with harsh working conditions or very stringent requirements for maintenance. However, the advance of active magnetic bearings in the last two decades has made them an increasingly mature and auspicious technology that can be applied to a wide range of high-speed rotating machines.

An overview of advantages and drawbacks of studied bearing types is given in Table 6.1.

References

1. J.F. Gieras, High speed machines, in *Advancements in Electric Machines (Power Systems)*, ed. by J.F. Gieras (Springer, Berlin 2008)
2. A. Binder, T. Schneider, High-speed inverter-fed ac drives, in *Electrical Machines and Power Electronics, 2007. ACEMP '07. International Aegean Conference on*, pp. 9–16, 10–12 Sept 2007
3. L. Wang, R.W. Snidle, L. Gu, Rolling contact silicon nitride bearing technology: a review of recent research. Wear **246**(1–2), 159–173 (2000)
4. E. Owen, Flexible shaft versus rigid shaft electric machines for petroleum and chemical plants, in *Petroleum and Chemical Industry Conference, 1989, Record of Conference Papers. Industrial Applications Society, 36th Annual*, pp. 157–165, 11–13 Sept 1989
5. J. Donaldson, High speed fans, in *High Speed Bearings for Electrical Machines (Digest No: 1997/164), IEE Colloquium on*, pp. 2/1–210, 25 Apr 1997
6. A. Dowers, The pursuit of higher rotational speeds; developments in bearing design and materials, in *High Speed Bearings for Electrical Machines (Digest No: 1997/164), IEE Colloquium on*, pp. 5/1–5/5, 25 Sept 1997
7. A. Muszynska, Whirl and whip-rotor/bearing stability problems. J. Sound Vib. **110**(3), 443–462 (1986)
8. S.H. Crandall, A heuristic explanation of journal bearing instability, in *Proceedings of the Workshop on Rotordynamic Instability Problems in High-Performance Turbomachinery, Texas A&M University, College Station, Texas*, pp. 274–283, 1982
9. Dyson Digital Motors, Dyson Ltd. (2009), http://www.dyson.com/technology/ddmtabbed.asp
10. C. Zwyssig, J.W. Kolar, S.D. Round, Megaspeed drive systems: pushing beyond 1 million r/min. IEEE/ASME Trans. Mechatron. **14**(5), 598–605 (2009)
11. C. Zwyssig, J. Kolar, W. Thaler, M. Vohrer, Design of a 100 W, 500000 rpm permanent-magnet generator for mesoscale gas turbines, in *Industry Applications Conference, 2005. Fourtieth IAS Annual Meeting. Conference Record of the 2005*, vol. 1, pp. 253–260, 2–6 Oct 2005
12. J. Peirs, D. Reynaerts, F. Verplaetsen, Development of an axial microturbine for a portable gas turbine generator. J. Micromech. Microeng. **13**(4), S190 (2003)
13. Radical Design Improvements with Hybrid Bearings in Electric Drives for Core Drill and Stone Saw Systems, SKF (2001), http://www.skf.com/files/001278.pdf
14. M. Weck, A. Koch, Spindle bearing systems for high-speed applications in machine tools. CIRP Ann. Manuf. Technol. **42**(1), 445–448 (1993)
15. L. Zheng, T. Wu, D. Acharya, K. Sundaram, J. Vaidya, L. Zhao, L. Zhou, K. Murty, C. Ham, N. Arakere, J. Kapat, L. Chow, Design of a super-high speed permanent magnet synchronous

motor for cryogenic applications, in *Electric Machines and Drives, 2005 IEEE International Conference on*, pp. 874–881, 15 May 2005

16. L. Burgmeier, M. Poursaba, Ceramic hybrid bearings in air-cycle machines. J. Eng. Gas Turbines Power **118**(1), 184–190 (1996)

17. H. Aramaki, Y. Shoda, Y. Morishita, T. Sawamoto, The performance of ball bearings with silicon nitride ceramic balls in high speed spindles for machine tools. J. Tribol. **110**(4), 693–698 (1988)

18. X. Luo, K. Cheng, D. Webb, F. Wardle, Design of ultraprecision machine tools with applications to manufacture of miniature and micro components. J. Mater. Process. Technol. **167**(2–3), 515–528 (2005)

19. D. Huo, K. Cheng, F. Wardle, Design of a five-axis ultra-precision micro-milling machine—UltraMill. Part 1: holistic design approach, design considerations and specifications. Int. J. Adv. Manuf. Technol. **47**(9–12), 867–877 (2009)

20. L.G. Frechette, S.A. Jacobson, K.S. Breuer, F.F. Ehrich, R. Ghodssi, R. Khanna, C.W. Wong, X. Zhang, M.A. Schmidt, A.H. Epstein, High-speed microfabricated silicon turbomachinery and fluid film bearings. J. Microelectromech. Syst. **14**(1), 141–152 (2005)

21. Air Bearing Application and Design Guide, New Way Precision (2003), http://www.newwayprecision.com

22. G. Genta, *Dynamics of Rotating Systems* (Springer, Berlin, 2005)

23. B.H. Rho, K.W. Kim, A study of the dynamic characteristics of synchronously controlled hydrodynamic journal bearings. Tribol. Int. **35**(5), 339–345 (2002)

24. A. Muszynska, Stability of whirl and whip in rotor/bearing systems. J. Sound Vib. **127**(1), 49–64 (1988)

25. B. Majumdar, Externally pressurized gas bearings: a review. Wear **62**(2), 299–314 (1980)

26. I. Pickup, D. Tipping, D. Hesmondhalgh, B. Al Zahawi, A 250,000 rpm drilling spindle using a permanent magnet motor, in *Proceedings of International Conference on Electrical Machines—ICEM'96*, pp. 337–342, 1996

27. T. Osamu, T. Akiyoshi, O.N.O. Kyosuke, Experimental study of whirl instability for externally pressurized air journal bearings. Bull. Jpn. Soc. Mech. Eng. **11**(43), 172–179 (1968)

28. C.-H. Chen, T.-H. Tsai, D.-W. Yang, Y. Kang, J.-H. Chen, The comparison in stability of rotor-aerostatic bearing system compensated by orifices and inherences. Tribol. Int. **43**(8), 1360–1373 (2010)

29. K. Czolczynski, K. Marynowski, Stability of symmetrical rotor supported in flexibly mounted, self-acting gas journal bearings. Wear **194**(1–2), 190–197 (1996)

30. K. Stout, F. Sweeney, Design of aerostatic flat pad bearings using pocketed orifice restrictors. Tribol. Int. **17**(4), 191–198 (1984)

31. R. Bassani, E. Ciulli, P. Forte, Pneumatic stability of the integral aerostatic bearing: comparison with other types of bearing. Tribol. Int. **22**(6), 363–374 (1989)

32. A. Mohamed, F. Emad, Nonlinear oscillations in magnetic bearing systems. IEEE Trans. Autom. Control **38**(8), 1242–1245 (1993)

33. G. Schweitzer, Active magnetic bearings—chances and limitations, in *6th International Conference on Rotor Dynamics*, pp. 1–14, 2002

34. M. Kimman, H. Langen, J. van Eijk, H. Polinder, Design of a novel miniature spindle concept with active magnetic bearings using the gyroscopic stiffening effect, in *Proceedings of the 10th International Symposium on Magnetic Bearing, Martigny, Switzerland*, 2006

35. R. Larsonneur, P. Bnhler, P. Richard, Active magnetic bearings and motor drive towards integration, in *Proceedings 8th International Symposium Magnetic Bearing, Mito, Japan*, pp. 187–192, 2002

36. M. Kimman, H. Langen, R.M. Schmidt, A miniature milling spindle with active magnetic bearings. Mechatronics **20**(2), 224–235 (2010)

37. M. Ahrens, L. Kucera, R. Larsonneur, Performance of a magnetically suspended flywheel energy storagedevice. IEEE Trans. Control Syst. Technol. **4**(5), 494–502 (1996)

38. H. Fujiwara, K. Ebina, N. Takahashi, O. Matsushita, Control of flexible rotors supported by active magnetic bearings, in *Proceedings of the 8th International Symposium on Magnetic Bearings, Mito, Japan*, 2002

39. G. Schweitzer, E. Maslen (eds.), *Magnetic Bearings: Theory, Design and Application to Rotating Machinery* (Springer, Berlin, 2009)
40. H. Balini, H. Koroglu, C. Scherer, Lpv control for synchronous disturbance attenuation in active magnetic bearings, in *ASME 2008 Dynamic Systems and Control Conference*, vol. 2008, no. 43352, pp. 1091–1098, 2008
41. W. Canders, N. Ueffing, U. Schrader-Hausman, R. Larsonneur, MTG400: a magnetically levitated 400 kW turbo generator system for natural gas expansion, in *Proceedings of the 4th International Symposium on Magnetic Bearings*, pp. 435–440, 1994
42. J. Schmied, Experience with magnetic bearings supporting a pipeline compressor, in *Proceedings of the 2nd International Symposium on Magnetic Bearings*, pp. 47–56, 1990
43. Y. Suyuan, Y. Guojun, S. Lei, X. Yang, Application and research of the active magnetic bearing in the nuclear power plant of high temperature reactor, in *Proceedings of the 10th International Symposium on Magnetic Bearings*, 2006
44. C.R. Knospe, Active magnetic bearings for machining applications. Control Eng. Pract. **15**(3), 307–313 (2007)
45. . J.-K. Park, S.-K. Ro, B.-S. Kim, J.-H. Kyung, W.-C. Shin, J.-S. Choi, A precision meso scale machine tools with air bearings for microfactory, in *5th International Workshop on Microfactories, Besancon, France*, 2006
46. . R. Hebner, J. Beno, A. Walls, Flywheel batteries come around again. IEEE Spectr. **39**(4), 46–51 (2002)
47. R. Blom, M. Kimman, H. Langen, P. van den Hof, R.M. Schmidt, Effect of miniaturization of magnetic bearing spindles for micro-milling on actuation and sensing bandwidths, in *Proceedings of the Euspen International Conference, EUSPEN 2008, Zurich, Switzerland*, 2008
48. T. Wang, F. Wang, H. Bai, H. Cui, Stiffness and critical speed calculation of magnetic bearing-rotor system based on fea, in *Electrical Machines and Systems, 2008. ICEMS 2008. International Conference on*, pp. 575–578, 17–20 Sept 2008
49. R. Humphris, P. Allaire, D. Lewis, L. Barrett, Diagnostic and control features with magnetic bearings, in *Energy Conversion Engineering Conference, IECEC-89, Proceedings of the 24th Intersociety*, vol. 3, pp. 1491–1498, 6–11 Aug 1989

Chapter 7
Design of the High-Speed-Spindle Motor

7.1 Introduction

As explained in Sects. 1.2 and 1.3 of the thesis Introduction, the PhD project started within the Dutch Microfactory framework with the goal of development of a built-in electrical spindle drive which would facilitate high rotational speed of the spindle and accurate micro-milling. This chapter will present the design of the spindle motor, from a conceptual design to electromagnetic and structural optimization of the motor.

In the time when the project started, some experience with machining spindles had already been gained within the Microfactory project and important limitations of high-speed spindles had been foreseen. This greatly influenced the design of the spindle drive. Namely, not only that the design aimed at improvements in electromagnetic actuation of the existing high-speed spindles, but it also looked for radically new spindle concepts which would overcome speed limits of existing spindles. As a result, a concept of a frictionless short-rotor spindle was born.

Influence of the analyses and models presented in Chaps. 2–6 on the spindle-motor design was twofold. Definition of speed limits of permanent magnet machines greatly affected initial, conceptual design of the new spindle drive. At the same time, the developed models formed an analytical basis for the motor design and optimization.

The Chapter will start with presenting development of new spindle concepts in the Microfactory project group—Sect. 7.2. Thereafter, geometric and electromagnetic design of the spindle motor will be explained in Sects. 7.3 and 7.4, then evaluated using FEM in Sect. 7.5 and, lastly, optimization of the rotor retaining sleeve is presented in Sect. 7.6.

7.2 New Spindle Concepts

In the Microfactory project a small spindle in active magnetic bearings (AMB) was realized by Kimman [1, 2]. A relatively slender rotor is supported by two radial bearings and an axial bearing which exerts force over a small rotor disc (Fig. 7.1).

A. Borisavljević, *Limits, Modeling and Design of High-Speed Permanent Magnet Machines*, 129
Springer Theses, DOI: 10.1007/978-3-642-33457-3_7,
© Springer-Verlag Berlin Heidelberg 2013

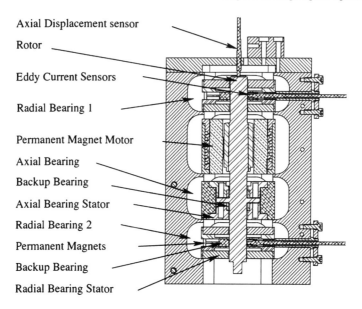

Axial Displacement sensor

Rotor

Eddy Current Sensors

Radial Bearing 1

Permanent Magnet Motor

Axial Bearing

Backup Bearing

Axial Bearing Stator

Radial Bearing 2

Permanent Magnets

Backup Bearing

Radial Bearing Stator

Fig. 7.1 Section view of the AMB spindle, taken from [1]

Rotation of the spindle is controlled by a commercially available PM motor. The maximum attained speed of the spindle is 150.000 rpm. It was shown that miniaturization of a spindle has positive effects on actuation and cutting force monitoring [1, 3].

The first flexural critical speed (approximately 180000 rpm) of the spindle has represented an obstacle of utilizing the motor up to its maximum speed constrained by the structural limit of the PM rotor (250000 rpm). Additionally, relatively high negative stiffness of the motor caused runout higher than it was expected [1].

It was apparent that, for reaching higher rotational speeds, the spindle length would need to be significantly decreased. However, that was hardly achievable with the same motor-bearings configuration. Hence, Kimman et al. [4] proposed a whole new approach for high-speed spindles: to use a short (disc-shaped) rotor suspended in AMB. The inspiration was found in an idea of 3DOF combined axial and radial magnetic bearings envisaged in [5]. Kimman et al. [4] proposed using such bearings for supporting 5DOF of a disc thus benefiting from reducing rotor tilting and higher resonance frequencies (Fig. 7.2). In essence, it would mean that all the bearings from the original setup—Fig. 7.1—would be grouped around the axial-bearing disc and that would, in turn, drastically reduce the spindle volume.

Advantages of using short rotor follow also from analyses of Chap. 5. A rigid short rotor has one critical speed less than its long/slender counterpart and it may also benefit from increase of critical speeds as a result of gyroscopic stiffening (see Fig. 5.12 in Sect. 5.5). Still, the greatest advantage of such a rotor clearly comes from the increase of flexural resonance frequencies as a result of the great reduction of the

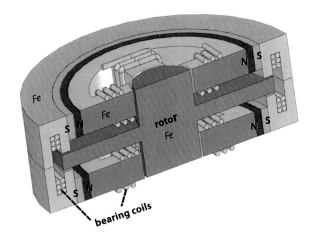

Fig. 7.2 5DOF magnetic bearings, as proposed by Kimman [6]

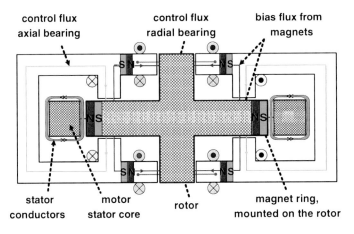

Fig. 7.3 Short-rotor spindle—concept ©2009 IEEE

rotor slenderness. In that way, stability threshold at the first flexural critical speed is too high to be reached and it ceases to be the limiting factor for the rotational speed.

The next step in the development of a short-rotor spindle was to integrate an electrical motor into the 5DOF bearings from Fig. 7.2. It was apparent that close spatial integration of AMB and motor was needed. Several concepts for spindle motor were considered including bearingless motors [7, 8] and axial-flux machines. It was a standard, radial-flux PM motor, however, that offered possibility of motor-bearings integration without merging their function and without changing the original AMB concept. The conceptual design of the new spindle is depicted in Figs. 7.3 and 7.4.

In proposed magnetic bearings (Fig. 7.3) control actions in axial and radial directions are decoupled: permanent magnets are utilized both to create bias flux and to separate control fluxes in different directions. Armature flux of the PM motor passes

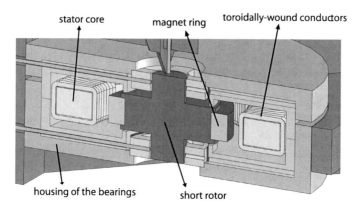

Fig. 7.4 Short-rotor spindle—a cross-section ©2009 IEEE

through the rotor disc, without interfering with the bearings' control fluxes. In that way, performance of the motor and the AMB are not simultaneously compromised in the design stage and their control is virtually independent during operation. A slotless stator of the motor is fitted into the axial bearings to facilitate good thermal contact with the environment. Toroidal windings represent a sensible solution for a motor with such a small stack length in comparison with the radius.

Finally, it was realized, after work on the spindle design started, that it would be convenient to have a setup with a passive bearing type for testing the electrical drive separately, without any coupling with magnetic bearings. For this purpose, aerostatic bearings were chosen for a number of reasons. They are relatively simple, frictionless and have high motion accuracy and stiffness. Simple orifices were chosen as air restrictors since they have not been associated with any pneumatic instability [9]. The concept of the air-bearings setup is presented in Figs. 7.5 and 7.6.

The aerostatic bearings for supporting the short rotor were developed by Tsigkourakos [10] under supervision of Kimman and the thesis' author. Two journal air bearings support the rotor in radial directions. The axial/thrust bearing is preloaded by a permanent magnet to provide high bearing stiffness. The PM motor shares again the same housing with the bearings—it was planned that, for the motor, magnetic- and air-bearings from Figs. 7.3 and 7.5 are interchangeable.

The air-bearings setup was used as a test setup for the motor in this thesis. More details on the setup are given in Sect. 9.2.

7.3 Conceptual Design of the Motor

At the time of the motor design many of the application requirements were still missing. Namely, the project group at TU Delft lacked practical experience with micromilling and information on torque requirements for high-speed micromilling was rather scarce. Additionally, the work on the motor design started long before

Fig. 7.5 Aerostatic bearings for the short rotor—concept [10]

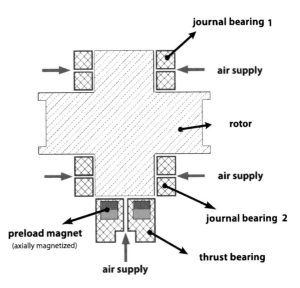

models presented in this thesis were developed. Therefore, most of the actual requirements were imposed by the designers themselves and not directly by the prospective application. However, the goal of the design was to offer a good proof of concept rather than a definite solution for micromilling spindles. Taking into consideration available data and technological limits in an academic environment, the designers looked for sensible and adequate requirements for the new spindle and the spindle motor in particular. They are presented as follows:

1. The motor needs to fit both proposed bearings and its stator should have a good thermal contact with the (bearings') housing.
2. It should be possible for both setups to be manufactured using technology of a standard/university workshop and of-the-shelf components. In other words, miniaturization of the rotor is limited by technological capabilities in the university environment. This is particularly important in the example of AMB whose extreme miniaturization would require utilization of advanced processing techniques and/or components (miniature sensors, coils, etc.).
3. In light of the previous requirement, it was concluded that a rotational speed of 200.000 rpm of a disc-shaped rotor in proposed bearings would be technically achievable (see also Sect. 7.3.1). Therefore, that speed was set as the speed requirement for the motor.
4. No data on required torque for micromilling were available. However, available estimations of cutting forces for milling with sub-millimeter tools suggested rather small load forces, in order of a few newtons and lower [1, 11]. It was, thus, expected that the load torque would be significantly smaller then the drag resulting from windage and eddy-currents. Taking into account worst predictions of air-friction loss (Fig. 3.21 in Sect. 3.6.3) it was concluded that 200 W of power would be certainly sufficient for operation at the required maximum speed.

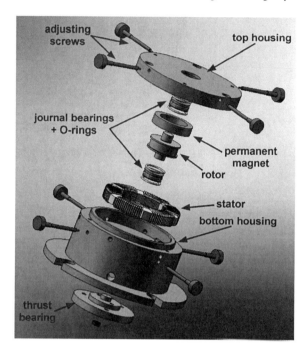

Fig. 7.6 Setup with the aerostatic bearings—exploded view [10]

5. For a high-frequency-operating machine which would be completely enclosed in magnetic bearings, minimization of frequency-dependant losses was extremely important. Besides, due to inability to reliably model air-friction loss and losses in the permanent magnet (Sects. 3.6.3 and 3.6.4), thermal model was not developed, thus, mitigation of losses was inevitable for a safe design. Therefore, minimum loss (= maximum efficiency) was taken as a decisive criterion for both component choice and electromagnetic design.
6. As pointed out in Sect. 3.5, negative motor stiffness, which results from the unbalanced magnetic force, must be, at least, an order of magnitude lower than the stiffness of the radial bearings and the unbalanced force must be lower than the bearing force capacity. The stiffness limit was critical for the case of AMB (estimated in order of 10^5 N/m) while the force limit was critical for the aerostatic bearings.
7. Structural robustness of the rotor was an equally important requirement for the design. Proper retaining of the magnet was crucial, particularly for a high-speed rotor with a high diameter to length ratio.

All these requirements affected the design of the motor whose conceptual design is depicted in drawings in Fig. 7.7 The motor concept is explained in the rest of this section (Fig. 7.8).

A laminated, slotless stator core has protrusions corresponding to the axial direction for good thermal contact with the housing. Advantages of slotless machines for very-high-speed operation were discussed in Chap. 2. Exclusion of stator teeth

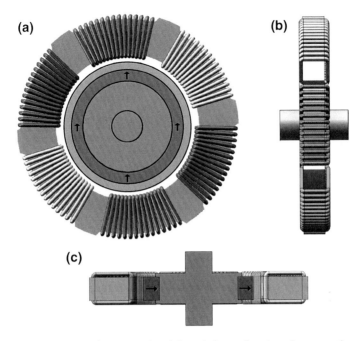

Fig. 7.7 Spindle motor drawing: **a** top view, **b** lateral view and **c** a lateral cross-section

removes slotting-effect harmonics from the PM field while, at the same time, reduces impact of armature-field harmonics in the PM rotor. As a whole, a slotless motor is prone to be more efficient and less susceptible to rotor overheating than its slotted counterpart.

Conductors are wound toroidally over the core, thus, dispensing with, for this case, unavoidably long end windings. The windings are non-overlapping, i.e. each phase winding is uniformly wound over two 60°-sections of the stator circumference.

A plastic-bonded magnet of the injection molded type is applied onto the incised part of the rotor disc (see Fig. 7.10). An incision is previously made in the shaft for a better fit of the magnet. The injection molded magnet contains very small magnet particles that are blended with a plastic binder—PPS. After applying this mixture onto the shaft at a very high temperature, the magnet will apply a stress on the shaft during the cooling in the mould. Such a magnet is very resistive to eddy-current losses. At the same time, low remanent flux density of the magnet, as a result of the plastic binder overtaking a great portion of the magnet volume, is quite adequate for a very-high-speed machine (see Sect. 2.4). The magnet is diametrically magnetized providing a perfectly sinusoidal back emf.

Finally, in order to sustain very strong centrifugal force at high speeds and ensure transfer of torque in the rotor throughout the whole range of speeds, the magnet needs to be contained in a non-magnetic enclosure/sleeve. A non-conductive sleeve has been conceived as a combination of glass and carbon fiber: details on the sleeve design are presented in Sect. 7.6.

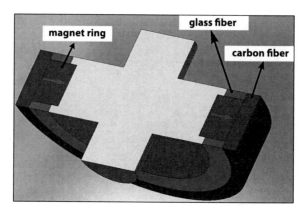

Fig. 7.8 Rotor design—concept

In the rest of the section the materials used for the motor parts—stator core, conductors, permanent magnet and sleeve—are discussed.

7.3.1 Stator Core

Resistivity to induced losses was a decisive factor in choosing the stator core material. Amorphous iron has excellent figures of losses and very favorable magnetic properties [12, 13] in comparison with other, more common core materials. However, amorphous iron is brittle and available only in form of ring tape cores and could not be processed to fit into the bearings.

Different types of silicon steel were considered. It is well known [14] that alloys with high, approximately 6.5 %, content of Si show excellent properties in terms of minimum induced loss and maximum permeability. However, steels with such a high Si content used to suffer from hardness and brittleness.

Workability of 6.5 % Si-steel has lately been improved [14] and such steel has been used in this project. The loss versus frequency characteristics of this steel are compared with amorphous iron and with high-frequency 0.12 mm Si-steel with a lower percentage of Si. The characteristics are presented in a graph (Fig. 7.9). Loss figures for Si-steel with 6.5 % Si are comparable to amorphous iron and significantly better than high-quality standard Si-steel.

7.3.2 Conductors

A drawback of the slotless type of PM machines is deteriorated cooling of the conductors in the air-gap. Namely, the absence of stator teeth makes transfer of heat

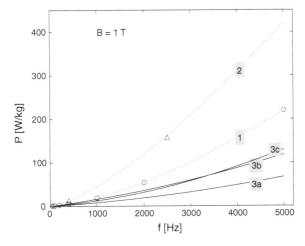

Fig. 7.9 Loss power density of different lamination materials versus frequency at 1 T flux density: 1—6.5 % *Si*-steel [15]; 2—0.12 mm Si-steel laminations [16]; 3—different samples of amorphous iron [13] ©2009 IEEE

from the air-gap conductors towards the stator core difficult [17]. In order to alleviate this effect, a special type of self-bonded wires is used. The wire [18] contains an adhesive surface varnish that interconnects wires after curing, enhancing the thermal conduction.

7.3.3 Permanent Magnet

As already mentioned, an injection-molded plastic-bonded magnet has been used in the rotor. This type of magnet was chosen for the following reasons:

- High resistivity: the rotor is barely cooled and preventing magnets from overheating is essential;
- Shape flexibility: injection-molded magnet could be directly applied onto the shaft in a ring form, keeping the high-speed rotor relatively well-balanced;
- A relatively small remanent field of the magnet (0.5 T) is adequate for the application.

However, very low yield stress, both compressive and tensile, is a great drawback of this magnet type and a lot of attention was given to structural designing of the rotor. High temperature polymer PPS, favorable for injection molding, has been chosen as the plastic binder. This binder type provides comparably the best mechanical characteristics among plastic-bonded magnet types [19, 20].

Fig. 7.10 Rotor shaft with
dimensions in mm

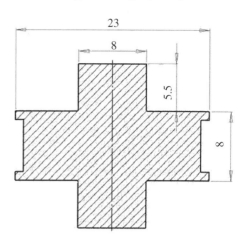

7.3.4 Magnet Retaining Sleeve

Carbon fibres were chosen as the enclosure material for their light weight and exceptional strength. The main drawback of using carbon fibres is their negligible thermal expansion in contrast with that of permanent magnets; therefore, additional stress on the enclosure is expected at elevated rotor temperatures. Furthermore, while able to withstand extreme tensions, carbon fibres are rather sensitive to bending [21] and they need to be protected from being cut at edges of neighbouring materials.

7.4 Motor Optimization

7.4.1 Rotor Shaft Design

Design of the rotor iron shaft was an initial step of the motor design since the rotor dimensions were also decisive for the bearing design. Length and diameter of the narrow shaft parts were determined first as a part of the conceptual design of the magnetic bearings. The length of 5 mm was estimated as a minimum required by the bearings to be available at both sides of the disc so that the bearings can be manufactured in the university workshop and can provide the desired stiffness. Consequently, the length of the narrow shaft part was set at 5.5 mm.

Diameter and length of the rotor disc were determined so that the polar inertia of the whole rotor is considerably higher than the transversal inertia so that the motor can benefit from rotor self-aligning. Another considered factor was eventual force density of the rotor—with the resulting area of the incised part of the disc a prospective required force density of the motor for the given power demand (200 W at 200.000 rpm) should not be too high. Force density estimation gives:

$$F_d < \frac{P}{2\pi \omega r^2 l_s} \qquad (7.1)$$

where r is the disc radius.

With the dimensions given in Fig. 7.10 the ratio between polar and transversal moments of inertia of the final rotor was going to be higher than 1.3. Additionally, with the given rotor dimensions, force density of the motor at the maximum power was going to be lower than $1.7\,kN/m^2$, which is more than acceptable when compared with corresponding values for commercially available PM machines.

7.4.2 Electromagnetic Optimization of the Motor Geometry

The motor optimization was carried out in two steps. In the first step, which will be explained here, the machine dimensions and number of stator-conductor turns were determined for a minimum total loss in the stator. In the second optimization step, reported in the next subsection, frequency-dependent copper losses are considered and the conductors are optimized.

As it was pointed out in Sect. 7.3, loss minimization was taken as the ultimate criterion for the motor design. However, it was not overall motor efficiency that was sought in the machine design, since that energy efficiency of the spindle drive was never a goal *per se*. The main intention of the design was, actually, to mitigate the overheating of the motor that would result from frequency-dependent losses. This criterion was partly taken as a safe option for the motor design since a good thermal model of the motor was not available.

In the rotor, the permanent magnet is the most susceptible to heat that would result from the losses. However, protection of the magnet from overheating was greatly taken care of in the conceptual design: plastic-bonded magnet material is highly resistive to eddy-current loss and slotless machine design additionally alleviates the influence of armature-field harmonics on the rotor.

It was also assumed that air friction does not have a great influence on the temperature inside the high-speed rotor for two reasons: turbulent air flow at high speeds greatly improves removal of the resulting heat at the rotor surface and, also, carbon fibers serve as a good thermal insulator for the rotor due to their very low thermal conductivity.

On the other hand, air friction greatly influences temperature in the air gap and, consequently, at the inner stator surface and air-gap conductors. However, at the time of the design, the author had neither a good representation of the air-friction loss nor a model of convection between the disc and stator. Although air friction has a huge impact on the overall drag torque (Sect. 3.6.3), how the air friction at the disc surface correlates with temperatures in the rotor and stator has remained undetermined.

Eventually, it was concluded that a relatively large friction loss is inevitable for the disc-shaped rotor so the air-friction loss was removed from the design focus. It was clear that without a thermal model of the machine the optimization goal of

Table 7.1 Optimization results: parameters of the motor geometry

Parameter	Symbol	Value
Stator stack length	l_s	6 mm
Magnet inner radius	r_m	10.5 mm
Magnet thickness	l_m	4 mm
Maximum sleeve thickness	l_e	2 mm
Winding area thickness	l_w	1.5 mm
Stator yoke thickness	l_y	8.8 mm
Number of phase turns	$2N$	88 (80[a])
Phase resistance	R	0.35 Ω[b]
Phase inductance	L	43 μH[b]
Flux linkage amplitude	ψ_{max}	1.72 mWb[b]
Rotor moment of inertia	J	3.6 mg·m^2

[a]value actually implemented in the motor
[b]model estimations based on the actually implemented number of turns

minimizing machine operating temperature was hardly going to be reached. A level of arbitrariness in the design was present and accepted from the beginning.

Minimization of the total electromagnetic loss in the stator was taken as the optimization goal. The objective function to be minimized was the following:

$$P_{loss} = P_{Fe} + P_{Cu,DC}, \tag{7.2}$$

where the stator core loss P_{Fe} is estimated using Eq. (3.95) from Sect. 3.6.1.

The copper loss was simply estimated as DC conduction loss for the given conductor dimensions, thus, excluding losses due to proximity- and skin-effect:

$$P_{Cu,DC} = I^2 \frac{l_{Cu}}{\sigma_{Cu} A_{Cu}}, \tag{7.3}$$

$$A_{Cu} = \frac{A_w}{6N} k_{fill} k_{Cu}, \tag{7.4}$$

where A_w is the cross-section of the winding region, k_{fill} is the estimated fill factor of rounded conductors and k_{Cu} is roughly estimated copper cross-section within a single conductor. Total conductor length l_{Cu} is given by Eq. (3.111).

Three motor parameters were taken as the design variables: thickness of the magnet ring l_m, the number of turns per phase $2N$ and the thickness of the stator core l_y. The optimization is performed using MATLAB Optimization Toolbox. Optimization constraints were defined as follows:

1. In order to obtain sinusoidal currents for the motor, it was planned to design a PWM inverter drive for the motor. To support the desired fundamental frequency of 3.33 kHz (200.000 rpm), very fast switching, in order of 100 kHz, was expected. The machine voltage is limited by the capability of a PWM inverter to perform

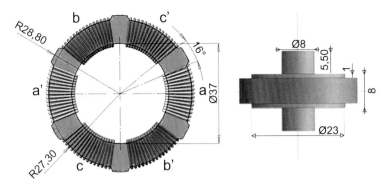

Fig. 7.11 Stator and rotor geometry with dimensions in mm

hard-switching at such a high frequency. It was therefore decided to limit the voltage of the machine to a value for which fast-switching MOSFETs are readily available. Amplitude of the no-load voltage at the maximum speed was set at 50 V:

$$\hat{e}\,(f = 3.33\,\text{kHz}) = 50\,\text{V} \tag{7.5}$$

For the optimization purpose the expression (3.47) for the no-load voltage was used.[1] With the given power reference (200 W) and the set value for the no-load voltage, the value of the stator current was practically determined:

$$I = \frac{P}{3Ek} \tag{7.6}$$

where k is a coefficient that accounts for the excess voltage over the phase impedance, arbitrarily set at 0.9.

2. The electromagnetic design was carried out before the modeling and design of the rotor carbon-fiber enclosure. The speed of sound in air was taken as the absolute limit for the rotor tip tangential speed which led to the maximum value for the outer radius of the rotor disc:

$$r_{e,\text{max}} \leq 16.5\,\text{mm.} \tag{7.7}$$

According to analysis presented in Sect. 4.2 on the stress in a rotating cylinder, the maximum rotation-influenced tangential stress in the carbon fiber sleeve with this outer radius would be smaller than:

---

[1] In the actual design, a mistake was made when the expression (3.47) was applied. Instead of setting 50 V as the no-load voltage amplitude, the desired rms value ($\frac{50}{\sqrt{2}} \approx 36$ V) is set for the voltage *amplitude*. Therefore, the actual winding flux linkage and power output of the test motor is lower than expected. It is important to notice, though, that the power requirement for the motor was set somewhat arbitrarily, based only on the projected frequency-dependent losses, and it is not crucial for the thesis.

$$\sigma_{cf,ref,\max} < \rho_{cf} v_t^2 = 190\,\text{MPa}, \tag{7.8}$$

which is still around 10 % of the maximum tensile stress in the fibers. This margin of stress consequently allows for additional stress in the fibers as a result of their fitting over the magnet.

3. In order to limit leakage of the magnet flux in the axial direction, the magnet thickness is set to comprise, at least, a half of the effective air gap:

$$l_e + g + l_w \le l_m \tag{7.9}$$

where l_e is the sleeve thickness, g is the mechanical air-gap and l_w is the thickness of the winding area.

4. The mechanical air gap is set at 0.5 mm
5. Eventually, an additional constraint of equalizing copper and iron losses is added, yielding somewhat balanced distribution of loss in the stator:

$$P_{Fe} = P_{Cu,DC}. \tag{7.10}$$

6. Naturally, saturation limit for the stator core is imposed. This limit hardly affected the optimization since saturation flux density in the core is, for the given magnet material, too high to be reached.

Optimization results are given in Table 7.1 and presented in drawings in Fig. 7.11.

7.4.3 Optimization of Conductors

Optimization in the previous subsection determined, practically, magnetic field in the machine, the available space for conductors and the number of conductor turns. In the second optimization step optimal conductors that fit the available winding space were selected so as to minimize copper loss for the estimated phase currents and the maximum current (electrical) frequency (3.33 kHz). The optimization procedure is quite similar to the one presented in [22].[2]

In order to alleviate skin- and proximity effect losses in a transformer or high-speed motor, parallel/stranded conductors are used. For very-high-frequency applications litz-wires are used which represent bundles of individually enameled strands. The bundle of strands is usually finally coated with cotton or silk. In this optimization step number and diameter of the strands within a phase conductor of the designed machine was defined.

In Sect. 3.6.2 losses in copper of a high-speed slotless machine were modeled. It was shown that skin effect has a very little influence on the losses in copper for reasonable speeds and conductor diameters of electrical machines. Therefore, only the DC conduction loss and eddy-current loss in the air-gap conductors were considered

---

[2] Most of the content of this section has been published in Borisavljevic et al. [23].

for the conductor optimization. Analytical expressions (3.104) and (3.106) for these two loss components can be, after considering Eqs. (3.107) and (3.110), rewritten here in the following form:

$$P_{Cu,DC} = \frac{k_1}{nd_{Cu}^2}, \tag{7.11}$$

$$P_{Cu,eddy} = k_2 nd_{Cu}^4, \tag{7.12}$$

where n and d_{Cu} are number and (copper) diameter of parallel conductors or strands within a single phase conductor and k_1 and k_2 are coefficients that depend on parameters which are determined in the previous subsection.

Hence, the total copper loss can be represented as:

$$P_{Cu} = P_{Cu,DC} + P_{Cu,eddy} = \frac{k_1}{nd_{Cu}^2} + k_2 nd_{Cu}^4. \tag{7.13}$$

For a given number of strands, an optimal strand diameter can be found so that the copper loss power is minimal (see Fig. 7.12):

$$\frac{\partial P_{Cu}}{\partial d_{Cu}}(d_{Cu,opt}) = -2\frac{k_1}{nd_{Cu,opt}^3} + 4k_2 nd_{Cu,opt}^3 = 0. \tag{7.14}$$

Thus, from (7.14) the optimal strand diameter yields:

$$d_{Cu,opt} = \frac{1}{\sqrt[3]{n}}\sqrt[6]{\frac{k_1}{2k_2}}, \tag{7.15}$$

where coefficients k_1 and k_2 stem directly from expressions for copper losses: (3.104), (3.110) and (3.106), (3.107), respectively:

$$k_1 = \frac{4I^2}{\pi\sigma_{Cu}}(2l_s + (r_{so} - r_s)\pi)6N, \tag{7.16}$$

$$k_2 = \frac{\pi\hat{B}_m^2\omega^2\sigma_{Cu}}{128}l_s 6N. \tag{7.17}$$

The expression (7.15) shows that the optimal diameter of copper within a strand decreases proportionally to the third root of the number of strands. After substituting expression (7.15) into Eq. (7.13), it can be seen that the same proportionality is valid also for the total copper loss:

$$P_{Cu} \sim \frac{1}{\sqrt[3]{n}}. \tag{7.18}$$

The trends of decrease of the optimal strand diameter and loss power for the machine optimized in the previous subsection is shown in Fig. 7.13.

With increasing number of strands and accordingly adjusted copper diameter, the minimum copper loss decreases. However, for an assigned number of turns this increase in number of strands is limited by available conductor area in the air gap. It is, thus, necessary to estimate the area comprised by the optimized conductors.

For a calculated copper diameter within a strand the total strand diameter including insulation build is found first. For wires in the range of $30 \div 60$ AWG[3] $(0.254 \div 0.009$ mm) the total strand diameter can be estimated as [24]:

$$d_{st,opt} = d_r \alpha \left(\frac{d_{Cu,opt}}{d_r} \right)^{\beta}.$$

(7.19)

Coefficients in Eq. (7.19) were adjusted to fit manufacturer's data [25]: $\alpha = 1.12$, $\beta = 0.97$ and $d_r = 0.079$ mm.

For thicker wires, the following approximation is used [24]:

$$d_{st,opt} = d_{Cu,opt} + 10^{\frac{X-AWG}{44.6}} \cdot Y,$$

(7.20)

where $X = 0.518$ and $Y = 0.0254$ mm.

If several strands are used within a conductor, they are usually twisted together to form a bundle. It was assumed that the conductors are bundled if at least 3 strands are used. Diameter of a bundle was calculated in the following way:

$$d_{bundle} = p_f d_{st,opt} \sqrt{n},$$

(7.21)

where the packing factor p_f has the following values [25]:

$$p_f = \begin{cases} 1, & n = 1, 2 \\ 1.25, & 3 \le n \le 12 \\ 1.26, & 13 \le n \le 18 \\ 1.27, & 19 \le n \le 25 \\ 1.28, & n > 25 \end{cases}$$

(7.22)

The thickness of coating adds between 0.03 and 0.04 mm to the diameter of the bundle [24]:

$$d_{bundle,tot} = d_{bundle} + 0.035 \text{ mm}.$$

(7.23)

---

[3] American Wire Gauge—a standard set of wire conductor sizes in the US; the AWG number can be converted into mm using the approximate equation: d [mm] $= 0.127 \times 92^{\frac{36-AWG}{39}}$.

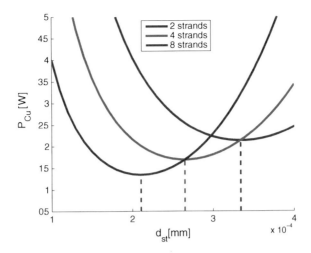

Fig. 7.12 Total copper loss versus strand diameter for different numbers of strands; *dashed lines* indicate optimal strand diameter

Finally the area required for the optimized conductors can be calculated as:

$$A_{w,opt} = \frac{1}{k_{fill}} \frac{d^2_{bundle,tot}}{4} \pi \cdot 6N, \qquad (7.24)$$

where $k_{fill} = 0.4$ is the fill factor of rounded conductors.

It is evident from Eqs. (7.15) and (7.21) that the increase of the bundle diameter with respect to number of strands is faster than corresponding decrease of diameter of individual strands. Therefore, the total area $A_{w,opt}$ will also increase with the number of strands. The limitation on the number of strands is eventually imposed by the available conductor area:

$$A_{w,opt} < A_w = \pi \left[r_s^2 - (r_s - l_w)^2 \right] \qquad (7.25)$$

Total required cross-sectional area of optimized conductors for the designed machine is plotted with respect to number of strands and compared with available conductor area in Fig. 7.14. For the designed motor only two parallel optimized conductors could fit the available area. Due to rather week magnet in the rotor, eddy-current losses in the air-gap conductors are not as pronounced as they would be in a slotless machine with high-energy magnets and that was decisive for such a low number of optimal parallel conductors.

Optimal diameter of the conductors was found to be 0.335 mm and the closest available diameter of Thermibond wires—0.314 mm—was used. Since only two parallel wires were needed, they were simply wound in parallel without additional twisting and coating.

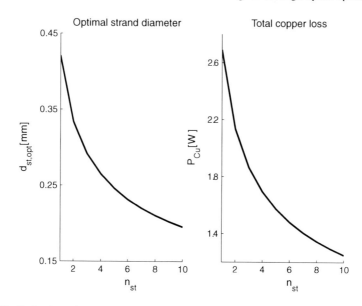

Fig. 7.13 Optimal conductor-strand diameter and corresponding total copper loss versus strand number

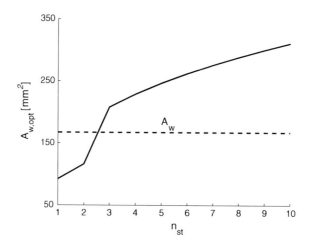

Fig. 7.14 Required area for optimized windings; available area is marked in *red*

7.5 FEM Design Evaluation

The motor electromagnetic design presented in the previous sections and some of the underlying models will be evaluated using finite-element simulations in this section. The FE models were built using Cedrat Flux2D/3D^{TM} software.

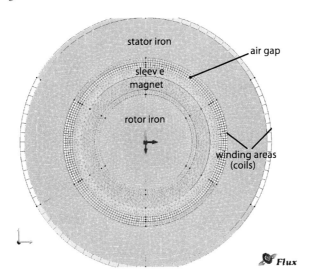

Fig. 7.15 Geometry and mesh of the 2D FE model of the designed motor

Table 7.2 2D FEM versus analytical model results

Parameter	Analytical	2D FEM
Amplitude of flux density in the windings (mT)	209.5	209.249
PM-flux linkage (mWb)	1.7285	1.7264
Phase inductance (μH)	43.427	64.02
Rotor loss (mW)	/	<1

7.5.1 2D FEM: Motor Parameters

The developed analytical models of the motor field have already been checked and confirmed by 2D FEM in Sects. 3.3.1 and 3.3.2. However, that FE model adopted the same assumptions as the 2D analytical model and, consequently, emulated the same physical model of the machine. The model does not take into account actual, toroidal distribution of the conductors since it assumes zero magnetic field outside the stator iron. External leakage of the armature field is, thus, neglected and the model geometry is equivalent to the geometry of a standard slotless machine with inserted air-gap conductors (see Fig. 3.4 in Chap. 3).

Here, results of a more-adequate 2D FEM of the test machine are presented in which the field is allowed to cross outside the motor external boundary. Calculations of those electromagnetic parameters of the motor that are directly derived from the magnetic field are compared to the results of the analytical model. The FEM geometry and mesh is given in Fig. 7.15 and results are shown in Table 7.2.

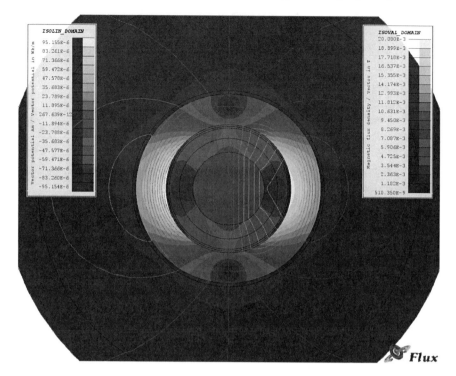

Fig. 7.16 Field lines and magnitude of the armature field modeled by 2D FEM

While the analytical representation of the permanent-magnet field is maintained—estimations of the PM flux density and the winding linkage of the PM flux of both models match very well—the total phase inductance of the motor seems to be significantly higher than analytically predicted. According to the FE model the leakage of the armature field outside the stator iron is significant (see Fig. 7.16) which is not surprising taking into account large effective air gap of the machine. This result shows inadequacy of the developed analytical model to represent the armature field in a toroidally-wound machine.

Rotor iron loss was modeled in a transient FE simulation for maximum expected current (2 A) and rotational frequency. The result partly confirms suitability of the magnetostatic approach in motor modeling since the losses in the rotor iron are extremely small. However, it was not possible to model eddy-current loss in the plastic-bonded magnet because no suitable means to represent the magnet electrical conductivity has been found. Still, the physical nature of such a magnet—permanent-magnet powder in a plastic binder—suggests that those losses are quite small.

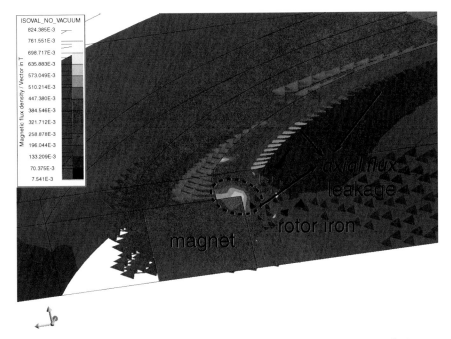

ISOVAL_NO_VACUUM

824.385E-3
761.551E-3
698.717E-3
635.883E-3
573.049E-3
510.214E-3
447.380E-3
384.546E-3
321.712E-3
258.878E-3
196.044E-3
133.209E-3
70.375E-3
7.541E-3

Magnetic flux density / Vector in T

axial flux leakage

rotor iron

magnet

Fig. 7.17 Zoom-view of the 3D FEM geometry: axial leakage of the PM flux

7.5.2 *3D FEM: No-Load Voltage and Phase Inductance*

To make a more realistic estimation of the motor parameters, a 3D FE model of the motor was created. Due to a rather short rotor and a very large gap between the magnet surface and stator inner surface, it was anticipated that leakage of the flux of the permanent magnet in the axial direction can have a significant influence on the motor performance. Furthermore, the effect of the external leakage of the armature field on the phase inductance was expected to be much higher than the 2D FE model suggested due to, beside else, a very short stator stack length.

From the simulation PM-flux linkage and, accordingly, no-load voltage appear to be noticeably lower (10 percent less) than predicted by the 2D analytical model which is a result of the axial flux leakage. The leakage is primarily enhanced by 1 mm extrusions of the rotor iron which support axial sides of the magnet—their presence partly short-circuits the field of the magnet (see Fig. 7.17).

Motor phase inductance is simulated in 3D FEM using non-meshed rectangular coils (it was not possible to create toroidal coils in the used software)—Fig. 7.18. The analytical model (and, also, 2D FEM) greatly underestimates the phase inductance: the analytical prediction is almost an order of magnitude lower than the prediction of the 3D FEM.

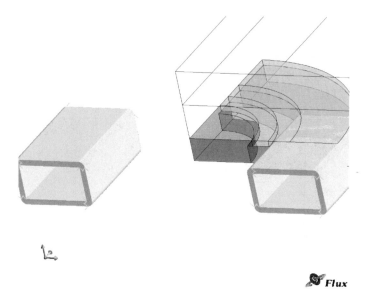

Fig. 7.18 One-eight of the motor 3D FEM and coil conductors of a single phase

7.5.3 2D FEM: Conductor Eddy-Current Loss

It would be extremely difficult to directly simulate losses in the conductors in FEM because of their disproportionally small size with respect to dimensions of the motor parts. Eddy-current loss in the conductor strands is, therefore, indirectly simulated. Firstly, the field in the center of the windings is calculated and simulated using 2D FEM. As already mentioned, 2D FEM upholds analytical field calculations and also affirms very low influence of the field of conductors on the total field in the winding area. It is also reasonable to assume that field of the eddy-currents do not have a noticeable influence on the PM field.[4]

Based on the calculation of the field in the winding, an abstract FE model of a thin solenoid is developed (Fig. 7.19) which creates the same flux density in the solenoid center as it is in the middle of the motor windings. Subsequently, a round conductor (strand) is added in the middle of the solenoid and, in a transient analysis, the losses in the conductor are calculated for different field frequencies and conductor diameters. Finally, the total eddy-current loss in the stator conductors is calculated by scaling the simulated loss with the total number of conductors (strands).

The eddy-current loss in the conductors calculated in this way perfectly matches analytical model predictions (Fig. 7.20). The FE simulations also confirm great importance of proper sizing of the air-gap conductors because eddy-current losses in conductors increase proportionally to the fourth power of the strand diameter (see Eq. (7.12)).

---

[4] A good part of this section has been published in Borisavljevic et al. [23].

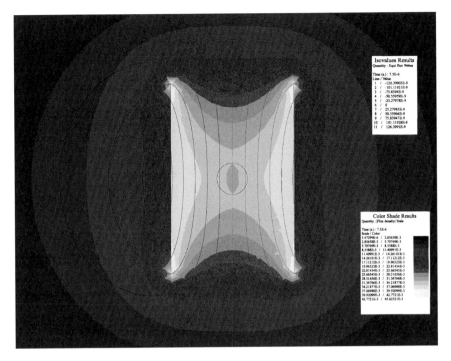

Fig. 7.19 A round conductor in a solenoid—finite element model

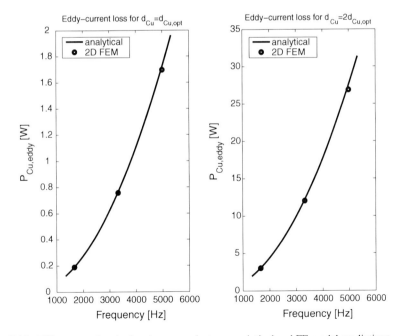

Fig. 7.20 Eddy-current loss in the air-gap conductors: analytical and FE model predictions

Table 7.3 PM-flux linkage and phase inductance according to analytical and FE models

	Analytical	2D FEM	3D FEM
PM-flux linkage (mWb)	1.7285	1.7264	1.5528
Phase inductance (μH)	43.4	64.0	350.0

7.6 Design of the Rotor Retaining Sleeve

Motor design represented in Sect. 7.4 determined electromagnetic properties of the motor. It is the magnet retaining sleeve that is supposed to ensure rotor structural integrity. Proper enclosing of the magnet became crucial for success of the design, particularly for a high-speed rotor with a high diameter to length ratio.[5]

Theoretical framework for structural modeling and design of a PM rotor is presented in Chap. 4 of this thesis. Particularly, in Sect. 4.4 an approach for the retaining sleeve optimization is developed; the approach was followed in the design of the enclosure of the test rotor and will be presented in this section.

7.6.1 Material Considerations: PPS-Bonded NdFeB Magnet

Injection-moulded bonded NdFeB magnets offer important advantages such as high resistivity to magnetically induced losses and shape flexibility, while their remanent field is sufficient for the application. Still, the possibility of plastic deformation and creep of the polymer material has remained a great concern.

High temperature polymer polyphenylene-sulfide (PPS), favourable for injection molding [19], was chosen as the plastic binder. Although PPS allows somewhat smaller volume fraction of permanent magnet material than some other polymer binders (e.g. nylon) and, accordingly, poorer magnetic properties [20], PPS-based bonded magnets have superior mechanical properties with respect to other bonded magnet types in terms of strength and processability [19, 20]. PPS also exhibits, for a polymer, high melting temperature (280 °C) and PPS-based magnets can maintain structural integrity even at the temperature of 180 °C [27]. Yet, at about 85 °C glass transition occurs in PPS [19] causing softening of the amorphous phase and a great reduction in strength [28]. Therefore, that temperature is taken as the rotor temperature limit (Table 7.3).

Scarce of data on mechanical properties of bonded magnets hampered the work on modeling of stress in the rotor. A valuable study on this issue was performed by Garell et al. [27] who offered data on tensile modulus and strength. However, the magnet endures compression from the enclosure and the properties must have been assessed for that condition. In the modeling it was assumed that the compressive

---

[5] Most of the content of this Section has been taken from Borisavljevic et al. [26].

modulus is equal to the tensile modulus. Further, compression was assumed to be limited by the compressive strength of PPS and the value of 120 MPa was imposed on the maximum stress in the magnet. Finally, Poisson's ratio and coefficient of thermal expansion (CTE) were estimated from the rules of mixtures [29]:

$$\nu_{bond} = (1 - p) \cdot \nu_{pps} + p \cdot \nu_{NdFeB}, \tag{7.26}$$

$$\alpha_{bond} = \frac{(1 - p) \cdot \alpha_{pps} E_{pps} + p \cdot \alpha_{NdFeB} E_{NdFeB}}{(1 - p) \cdot E_{pps} + p \cdot E_{NdFeB}}, \tag{7.27}$$

where p (=0.6) is the volume fraction of permanent magnet material in the bond and E_{pps} and E_{NdFeB} are tensile module of PPS and NdFeB magnet respectively.

7.6.2 Material Considerations: Carbon-Fiber Composite

Carbon fibers exhibit strong orthotropic nature since the fibers' mechanical properties differ considerably in different directions and strongly depend on the cross-sectional type [30]. This makes precise analytical modelling burdensome. Analyses in Chap. 4, however, showed that mechanical stress in the rotor can be quite adequately represented by assuming the composite properties in the direction of fibers (Table 7.4: || fibers) be valid in all directions.

7.6.3 Sleeve Optimization

Goal of the optimization of the rotor enclosure/sleeve was maximizing the rotational speed that the rotor could withstand. Maximum permissible thickness of the sleeve that could be fit in the air-gap is 2 mm. Temperature of the PPS glass-transition $-85\,°C$, $\Delta T_{max} \approx 60\,°C$—is set as the rotor temperature limit.

As pointed out is Chap. 4, the thickness of the enclosure and interference fit must be adequately chosen so that the contact pressure between the rotor parts is maintained and that the equivalent stress in each rotor part must be below the material ultimate stress (Eqs. (4.41) and (4.42)) throughout the whole speed range at the operating temperature.

Critical stresses in the rotor—are radial (contact) stress at the magnet-iron boundary and tangential stress (tension) at the enclosure inner surface—take on the following forms (Eqs. (4.43) and (4.44)):

$$\sigma^m_{r,crit} = F_1 (r_e) \cdot \Omega^2 - G_1 (r_e) \cdot \delta - H_1 (r_e) \cdot \Delta T, \tag{7.28}$$

$$\sigma^e_{\theta,crit} = F_2 (r_e) \cdot \Omega^2 + G_2 (r_e) \cdot \delta + H_2 (r_e) \cdot \Delta T, \tag{7.29}$$

Table 7.4 Properties of the carbon fibres

Material property	∥ Fibers	⊥ Fibers
Density (g/cm^3)	1.55	
Elastic modulus (GPa)	186.2	9.5
Maximum stress[a] (MPa)	1400	100
Poisson's ratio	$v_{12} = 0.3$	
CTE (μm/m/$^\circ$C)	-1	54

[a]The values assumed a great safety margin

Table 7.5 Properties of the bonded magnet and its constituents

Material property	PPS	NdFeB	Bond[a]
Density (g/cm^3)	1.35–1.7	7.35–7.6	4.8[b]
Compressive modulus (GPa)	2.8–3	140–170	31.7[c]
Compressive strength (MPa)	125–185	800–1100	120
Poisson's ratio	0.36–0.4	0.24	0.3[d]
CTE (μm/m/$^\circ$C)	40–55	4–8	4.7[d]

[a]Values used in modeling
[b]Measured
[c]Taken from [27]
[d]Calculated using the rules of mixtures

where $F_{1,2}$, $G_{1,2}$ and $H_{1,2}$ are positive functions of the sleeve outer radius r_e.

The optimal fit and theoretical maximum speed are obtained as functions of the sleeve radius r_e from the following system of equations:

$$\sigma_{r,crit}^{m}\left(\Omega = \Omega_{max}, \delta = \delta_{opt}, \Delta T = 0\right) = 0$$
$$\sigma_{\theta,crit}^{e}\left(\Omega = \Omega_{max}, \delta = \delta_{opt}, \Delta T = 60\,^\circ C\right) = \sigma_{max}^{e}. \tag{7.30}$$

For the optimal interference fit, both critical stresses are reached at a same maximum rotational speed which is, in turn, determined by the enclosure radius.

As it was expected, the highest speed could be reached with the highest allowable enclosure thickness of 2 mm. The optimal interference fit is obtained from (7.30) to be 110 μm.

Finally, in this design, additional limit is forced on the fitting by the magnet compression strength. Compliance of the calculated interference fit to that limit needs to be ensured. Namely, because of the polymer used in the magnet, compression strength of the bonded magnet is much lower than that of sintered magnets (Table 7.5), maximum amount of a press-fit that can be applied on the magnet is expected to be rather low.

To account for this, maximum permissible fit δ_{max} is calculated from the equation:

$$\sigma_{eq,crit}^{m}\left(\Omega = 0, \Delta T = 0\right) = \sigma_{max}^{m}, \tag{7.31}$$

where $\sigma_{eq,crit}^{m}$ is *von Mises* stress at the magnet inner surface:

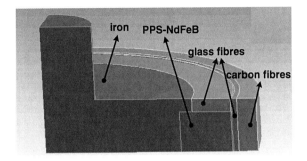

Fig. 7.21 Final rotor structure

$$\sigma^m_{eq,crit} = \sqrt{\sigma^{m2}_{r,crit} + \sigma^{m2}_{\theta,crit} - \sigma^m_{r,crit}\sigma^m_{\theta,crit}} \qquad (7.32)$$

and σ^m_{max} is assumed compression strength of the magnet.

From (7.31) maximum interference fit is calculated to be 95 μm and, since it is smaller than the previously calculated optimal fit, it is set as the definite value of the interference fit between the magnet and enclosure.

7.6.4 Final Design

Although it is a commonly applied technique in high-speed rotors, the test application rotor could not be simply enclosed by pressing a fiber ring over a bare magnet without causing its damage. Polymer magnet would also buckle if it was simply pressed by the carbon fibers.

Therefore, the rotor structure was designed so as to ensure structural integrity of both magnet and fibers. A quarter of a cross-section of the final rotor is given in Fig. 7.21.

In the final design glass fibres were used to enable safe pressing of the carbon fibre ring and to protect carbon fibres from bending at corners of the magnet.

The procedure for retaining the magnet was as follows: First the glass fibre rings are pressed on the shaft over top and bottom faces of the magnet. Outer surfaces of the magnet and the glass fibre rings are then polished before pressing of the enclosure. At the same time, carbon fibres are wound around a very thin glass fibre ring whose inner radius is for 95 μm smaller then the outer radius of the magnet. Finally, the carbon/glass fibre ring is pressed onto the rotor.

The rotor final structure was modeled in 3D using Ansys Workbench software. Compression at the magnet inner surface resulting only from the press-fit is much smaller than calculated from the 2D modelling. However, what concerns is a large stress concentration at the line close to boundary between the iron shaft and glass fibre ring (pointed by arrows at Fig. 7.22).

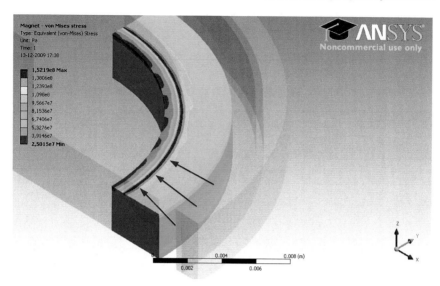

Fig. 7.22 Stress in the magnet as result of the sleeve fitting

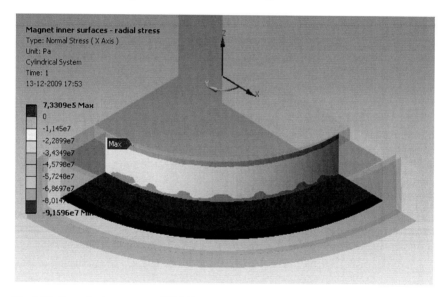

Fig. 7.23 Stress in the magnet at 180.000 rpm, room temperature

Further, according to this model, if the rotor remained at room temperature, the contact between the magnet and iron would be lost beyond 180.000 rpm (Fig. 7.23). Maximum possible speed is increased, though, if the operating temperature rises.

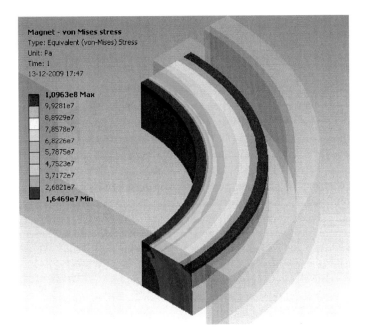

Fig. 7.24 Stress in the magnet at 200.000 rpm, 85 °C

Maximum equivalent stress in the magnet at the maximum speed and temperature is at the outer magnet surface and amounts to 110 MPa (Fig. 7.24) which is still below the compression limit.

Maximum tensile stress in carbon fibres—1147 MPa—is in a very good agreement with results from 2D modelling.

7.7 Conclusions

The chapter presents the design of the high-speed spindle motor, from a conceptual design to electromagnetic and structural optimization of the motor.

Two new spindle concepts[6] are presented; in both concepts a slotless toroidally-wound PM motor with a short rotor is combined with 5DOF frictionless—active magnetic or aerostatic—bearings. The motor is spatially integrated with bearings without merging their functions even in the example of active magnetic bearings. The spindle with a short rotor benefits from rotordynamical advantages of such rotors; most importantly, stability threshold becomes too high to be reached and the dynamical stability problems are simply avoided.

---

[6] Developed together with Kimman [2].

Loss minimization is taken as the ultimate criterion for the motor electromagnetic design. More precisely, the main intention of the design was to mitigate the overheating of the motor that would result from frequency-dependent losses. Chosen materials, conceptual design and electromagnetic optimization resulted in a relatively simple machine with very low electromagnetic losses.

Design of the rotor shaft was the initial step of the motor design since the rotor dimensions were also decisive for the bearing design. Diameter and length of the rotor disc were determined so that the polar inertia of the whole rotor is considerably higher than the transversal inertia so that the motor can benefit from rotor self-aligning.

The electromagnetic optimization was carried out in two steps. In the first step the machine geometry and number of stator-conductor turns were determined for a minimum total loss in the stator. In the second step the stator conductors are optimized for the given number of turns and machine loading defined in the preceding step. Exact analytical formulas for the optimization of air-gap conductors in slotless PM machines are derived which is an originality of this thesis.

The electromagnetic design of the motor is evaluated by FEM. The model of the PM field is confirmed by magnetostatic 2D FEM simulations and the field reduction influenced by the axial flux leakage is observed in 3D FE simulations. This leakage is enhanced due to the presence of the small extrusions of the rotor iron shaft which protect the magnet.

The armature field model developed in the thesis is not fully suitable to represent a motor with toroidal windings. The model does not take into account the actual, toroidal distribution of the conductors since it neglects the magnetic field outside the stator iron. However, as seen in the FE simulations and the external leakage of the armature field is immense, causing the motor phase inductance to be almost an order of magnitude higher than analytically predicted.

Eddy-current losses in motor air-gap conductors are indirectly simulated using an abstract FE model of a thin solenoid which creates the same flux density in the solenoid center as it is in the middle of the motor windings. The eddy-current loss in the conductors calculated in this way perfectly matches analytical model prediction and confirms great importance of proper sizing of the air-gap conductors.

Finally, transient FE models showed a negligible influence of eddy-current losses in the rotor iron on machine performance.

The rotor retaining sleeve is designed using the optimization approach described in Chap. 4. An innovative final design of a retaining sleeve for a short PM rotor is presented; the design consists of a combination of glass- and carbon-fiber retaining rings.

References

1. M. Kimman, H. Langen, R.M. Schmidt, A miniature milling spindle with active magnetic bearings. Mechatronics **20**(2), 224–235 (2010)
2. M.H. Kimman, *Design of a Micro Milling Setup with an Active Magnetic Bearing Spindle.* Ph.D. Dissertation, Delft University of Technology, 2010

3. R. Blom, M. Kimman, H. Langen, P. van den Hof, R.M. Schmidt, Effect of miniaturization of magnetic bearing spindles for micro-milling on actuation and sensing bandwidths, in *Proceedings of the Euspen International Conference, EUSPEN 2008, Zurich, Switzerland,* 2008

4. M. Kimman, H. Langen, J. van Eijk, H. Polinder, Design of a novel miniature spindle concept with active magnetic bearings using the gyroscopic stiffening effect, in *Proceedings of the 10th International Symposium on Magnetic Bearing, Martigny, Switzerland,* 2006

5. W. Canders, H. May, J. Hoffmann, Contactless magnetic bearings for flywheel energy storage systems, in *Proceedings of the 8th International Symposium on Magnetic Suspension Technology,* 2005

6. M. Kimman, H. Langen, J. van Eijk, R. Schmidt, Design and realization of a miniature spindle test setup with active magnetic bearings, in *Advanced Intelligent Mechatronics, 2007 IEEE/ASME International Conference on,* pp. 1–6, 4–7 Sept 2007

7. A. Chiba, T. Deido, T. Fukao, M. Rahman, An analysis of bearingless ac motors. IEEE Trans. Energy Convers. **9**(1), 61–68 (Mar. 1994)

8. M. Ooshima, A. Chiba, T. Fukao, M. Rahman, Design and analysis of permanent magnet-type bearingless motors. IEEE Trans. Ind. Electron. **43**(2), 292–299 (1996)

9. A. van Beek, *Machine Lifetime Performance and Reliability.* Delft University of Technology, 2004

10. P. Tsigkourakos, *Design of a Miniature High-Speed Spindle Test Setup Including Power Electronic Converter.* Master's Thesis, Delft University of Technology, 2008

11. R. Blom, P. van den Hof, Estimating cutting forces in micromilling by input estimation from closed-loop data, in *Proceedings of the 17th World Congress. The International Federation of Automatic Control, Seoul, Korea,* 2008

12. C. Zwyssig, J. Kolar, W. Thaler, M. Vohrer, Design of a 100 W, 500000 rpm permanent-magnet generator for mesoscale gas turbines, in *Industry Applications Conference, 2005. Fourtieth IAS Annual Meeting. Conference Record of the 2005,* vol. 1, pp. 253–260, 2–6 Oct 2005

13. W. Yuan, F. Liu, S. Pang, Y. Song, T. Zhang, Core loss characteristics of fe-based amorphous alloys. Intermetallics **17**(4), 278–280 (2009)

14. T. Yamaji, M. Abe, Y. Takada, K. Okada, T. Hiratani, Magnetic properties and workability of 6.5

15. JNEX-Core 10JNEX900, JFE Steel Corporation (2003), http://www.jfe-steel.co.jp/en/products/list.html

16. NO12-Core, Cogent Power Ltd. (2010), http://www.sura.se/Sura/hp_main.nsf/startupFrameset?ReadForm

17. N. Bianchi, S. Bolognani, F. Luise, High speed drive using a slotless pm motor. IEEE Trans. Power Electron. **21**(4), 1083–1090 (2006)

18. Enamelled wires: Thermibond® 158, Von Roll Isola (2004), http://products.vonroll.com/web/download.cfm?prd_id=1485&are_id=2&lng_id=EN

19. W. Sattich, J. Geibel, Reinforced poly(phenylene sulfide), in *Engineering plastics Handbook,* ed. by J.M. Margolis. (McGraw-Hill Companies, New York, 2006), pp. 385–418

20. X.L. Zhang, M.Y. Zhu, Y. Li, Q.P. Yang, H.M. Jin, J. Jiang, Y. Tian, Y. Luo, Study on fabrication process of anisotropic injection bonded nd-fe-b magnets. J. Iron Steel Res. Int. **13**(Suppl. 1, 0), 286–288 (2006), http://www.sciencedirect.com/science/article/pii/S1006706X0860196X

21. A. Binder, T. Schneider, M. Klohr, Fixation of buried and surface-mounted magnets in high-speed permanent-magnet synchronous machines. IEEE Trans. Ind. Appl. **42**(4), 1031–1037 (2006)

22. C. Zwyssig, J. Kolar, W. Thaler, M. Vohrer, Design of a 100 W, 500000 rpm permanent-magnet generator for mesoscale gas turbines, in *Industry Applications Conference, 2005. Fourtieth IAS Annual Meeting. Conference Record of the 2005,* vol. 1, pp. 253–260, 2–6 Oct 2005

23. A. Borisavljevic, H. Polinder, J.A. Ferreira, Conductor optimization for slotless PM machines, in *Proceedings of the XV International Symposium on Electromagnetic Fields in Mechatronics, ISEF 2011,* 2011

24. C. Sullivan, Optimal choice for number of strands in a litz-wire transformer winding. IEEE Trans. Power Electron. **14**(2), 283–291 (1999)
25. Enamelled Wires, Elektrisola, http://www.elektrisola.com/enamelled-wire.html
26. A. Borisavljevic, H. Polinder, J.A. Ferreira, Enclosure design for a high-speed permanent magnet rotor, in *Proceeding of the Power Electronics, Machines and Drives Conference, PEMD 2010*, 2010
27. M.G. Garrell, B.-M. Ma, A.J. Shih, E. Lara-Curzio, R.O. Scattergood, Mechanical properties of polyphenylene-sulfide (pps) bonded nd-fe-b permanent magnets. Mater. Sci. Eng. A **359**(1–2), 375–383 (2003)
28. Ryton PPS Data Sheets, Chevron Phillips Chemical Company (2000), http://www.cpchem.com/bl/rytonpps/en-us/Pages/RytonPPSDataSheets.aspx
29. L. Kollar, G. Springer, *Mechanics of Composite Structures* (Cambridge University Press, Cambridge, 2003)
30. H.-B. Shim, M.-K. Seo, S.-J. Park, Thermal conductivity and mechanical properties of various cross-section types carbon fiber-reinforced composites. J. Mater. Sci. **37**, 1881–1885 (2002)

Chapter 8
Control of the Synchronous PM Motor

8.1 Introduction

Permanent magnet machines have become prevalent among very high-speed machines and, to the author's knowledge, no machine other than permanent magnet has been reported to operate beyond the speed of 130.000 rpm. So-called PM synchronous machines (PMSM) with sinusoidal phase currents are generally preferred when low loss and smooth torque are important [1]. On the other hand, using high-frequency sinusoidal currents in the stator makes the design of a power converter more difficult and also complicates the implementation of control algorithm given a short switching period. A control method for PMSM must ensure stable, synchronous operation of the machine at high speeds having, at the same time, computational complexity tolerable by the given microcontroller. Since small high-speed machines, as a rule, lack space for a position/speed sensor, sensorless operation is usually required.[1]

Unlike induction machines, PM machines become unstable or marginally stable beyond a certain frequency when driven in open-loop [3]. Namely, due to the virtual absence of a component of the machine's torque which is proportional to the rotor speed, PMSM has a pair of poorly or negatively damped *rotor* poles [3, 4] and the machine is prone to lose synchronism when subjected to disturbance. Large synchronous machines have damper windings that suppress oscillations of speed during any transition by producing stabilizing *slip* torque like in an induction machine. However, having damper windings on a high-speed rotor is beyond consideration and necessary damping has to be introduced in a feedback control.

For achieving high performance, PM synchronous machines without damper windings are controlled using vector/field-oriented control. In the sensorless variant of the vector control, the position of the rotor is estimated by processing measured phase currents [5]. Nevertheless, sensorless vector control methods require a lot of computation while the processing power of standard microcontrollers and

---

[1] Most of the content of this chapter has been taken from Borisavljevic et al. [2].

A. Borisavljević, *Limits, Modeling and Design of High-Speed Permanent Magnet Machines*, 161
Springer Theses, DOI: 10.1007/978-3-642-33457-3_8,
© Springer-Verlag Berlin Heidelberg 2013

DSPs constrain their complexity when sample periods become very low. Therefore, very powerful or even dual DSPs are needed for realization of sensorless, close-loop control [6] and their price and complexity outweigh benefit that they bring [7].

Fortunately, applications where high-speed PMSM are used usually require no high performance of the speed control so that open-loop controllers can be used [4, 6, 8]. The open-loop control method that is mostly used for high-speed PMSM is V/f control with frequency modulation. In its basis, the method resembles V/f control of an induction motor: applied voltage is increased proportionally to frequency. In order to achieve stability, however, the reference frequency is modulated using perturbations of the active power and this frequency modulation, in effect, introduces necessary damping to the speed response. The method has been used regularly in open-loop control of high-speed PM machines [4, 8–10]; moreover, Itoh et al. [9] showed that, at high speeds, performance of this method is as good as or even better than performance of the sensorless vector control.

Voltage reference in the V/f control can be *a priori* set with respect to frequency, knowing machine parameters and the load torque. However, voltage is then applied irrespective of the actual current and load torque which may lead to overcurrents or, conversely, stalling and loss of synchronism. Dynamics of speed regulation in such a controller is very slow and only limited speed ramps can be supported [11].

Reference voltage is often actively calculated using measurements of the currents [4, 10]. Setting the voltage in this manner still has disadvantages: the voltage reference is strongly dependent on the machine's parameters and the calculation of the voltage involves a lot of demanding computations. Furthermore, such methods have been reported for rather low-speed machines [4].

Another possibility for adjusting the reference voltage to an optimum value is to use *efficiency control* [9] in which an *a priori* set value of reference voltage is modified during operation so as to achieve the d-axis current practically equal to zero.

Authors of [11] also used measurements of the reactive power but for perturbation of both reference voltage and frequency. Their solution offers faster dynamics of the speed control with respect to controllers in which only reference frequency is perturbated. Additionally, the reliance of the voltage-reference setting on machine parameters is not serious—it primarily depends on estimation of the phase inductance [11].

In the I/f control method a reference current is set instead of the voltage and the command voltages are generated through a current regulator [12, 13]. In papers in which the method has been reported the command current was set somewhat empirically and the method was used only for the start-up and low-speed range to prevent high current deviations in those operating regions; the control would be subsequently handed over to the standard V/f controller [12, 13].

In the thesis, the I/f method is used as a basis for stable, sensorless control of a high-speed PMSM. The method was developed for control of the spindle drive whose design is presented in the previous chapter. Stabilizing control (frequency modulation) from the standard V/f approach has been incorporated into the I/f method.

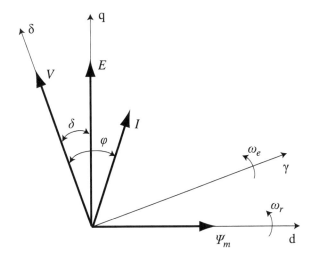

Fig. 8.1 Rotating reference frames, voltage control ©2010 IEEE

A computationally simple controller with an inherent control of the current (torque) has been developed. The controller was successfully tested in the practical setup.

This chapter starts with state-space modeling of an open-loop controlled PM machine and its stability is analyzed, Sect. 8.2. The V/f control method with modulation of the reference frequency is then presented in Sect. 8.3. The I/f control method, which is ultimately used for control of the spindle motor, is explained in Sect. 8.4. Finally, the control method is implemented in DSP and tested in the practical setup; results of the experiments are presented in Sect. 8.5.

8.2 Stability Analysis

Two rotating reference frames will be used to represent the electromagnetic quantities of the test motor: dq reference frame with the d-axis coinciding with the direction of the permanent magnet flux ψ_m and $\gamma\delta$ as controller's reference frame. In order to analyze open-loop stability of the rotor, δ-axis will be connected to the voltage phasor of the motor (as in the V/f control, Fig. 8.1). The reference frames rotate with electrical and mechanical angular frequency ω_e and ω_r, respectively, and the *power angle* δ also represents angular displacement between the frames.

State-space electromechanical model of the motor in $\gamma\delta$ reference frame is given in the following equations:

$$\frac{d}{dt}i_\gamma = -\frac{R}{L}i_\gamma + \omega_e i_\delta - \frac{\psi_m}{L}\omega_r \sin\delta + \frac{1}{L}v_\gamma$$
$$\frac{d}{dt}i_\delta = -\frac{R}{L}i_\delta - \omega_e i_\gamma - \frac{\psi_m}{L}\omega_r \cos\delta + \frac{1}{L}v_\delta$$
$$\frac{d}{dt}\omega_r = -\frac{3}{2}\frac{\psi_m}{J}i_\gamma \sin\delta + \frac{3}{2}\frac{\psi_m}{J}i_\delta \cos\delta - \frac{T_{load}(\omega_r)}{J} \qquad (8.1)$$
$$\frac{d}{dt}\delta = \omega_e - \omega_r$$

In (8.1) R and L denote phase resistance and inductance respectively, while i and v represent instantaneous current and voltage in the corresponding reference frame. After linearization of Eq. (8.1), the small-signal model of the motor is obtained:

$$\begin{bmatrix} \Delta \dot{i}_\gamma \\ \Delta \dot{i}_\delta \\ \Delta \dot{\omega}_r \\ \Delta \dot{\delta} \end{bmatrix} = \underline{A} \cdot \begin{bmatrix} \Delta i_\gamma \\ \Delta i_\delta \\ \Delta \omega_r \\ \Delta \delta \end{bmatrix} + \underline{B} \cdot \begin{bmatrix} \Delta v_\gamma \\ \Delta v_\delta \\ \Delta \omega_e \end{bmatrix}, \qquad (8.2)$$

where:

$$\underline{A} = \begin{bmatrix} -\frac{R}{L} & \omega_0 & -\frac{\psi_m}{L}\sin\delta_0 & -\frac{\psi_m}{L}\omega_0 \cos\delta_0 \\ -\omega_0 & -\frac{R}{L} & -\frac{\psi_m}{L}\cos\delta_0 & \frac{\psi_m}{L}\omega_0 \sin\delta_0 \\ \frac{3}{2}\frac{\psi_m}{J}\sin\delta_0 & \frac{3}{2}\frac{\psi_m}{J}\cos\delta_0 & -\frac{k_{L1}}{J} & \frac{3}{2}\frac{\psi_m}{J}\left(I_{\gamma 0}\cos\delta_0 - I_{\delta 0}\sin\delta_0\right) \\ 0 & 0 & -1 & 0 \end{bmatrix}, \qquad (8.3)$$

$$\underline{B} = \begin{bmatrix} \frac{1}{L} & 0 & I_{\delta 0} \\ 0 & \frac{1}{L} & -I_{\gamma 0} \\ 0 & 0 & 0 \\ 0 & 0 & 1 \end{bmatrix}. \qquad (8.4)$$

In (8.3) and (8.4) index 0 denotes steady-state value of the given variable and $\omega_{r0} = \omega_{e0} = \omega_0$ is the synchronous frequency. It is assumed that load torque T_{load} linearly depends on the rotational speed, i.e.:

$$T_{load}(\omega_r) = k_{T1}\cdot\omega_r + k_{T0}. \qquad (8.5)$$

Correlations between steady-state quantities are also derived in linearization of (8.1). Equilibrium between the electromagnetic and the load torque on the motor shaft yields the following equation:

$$\frac{3}{2}\psi_m I_{q0} - k_{T1}\omega_0 - k_{T0} = 0 \qquad (8.6)$$

In further analysis it will be assumed that the steady-state d current is zero. Hence, the expressions for components of the steady-state currents in $\gamma\delta$ reference frame are:

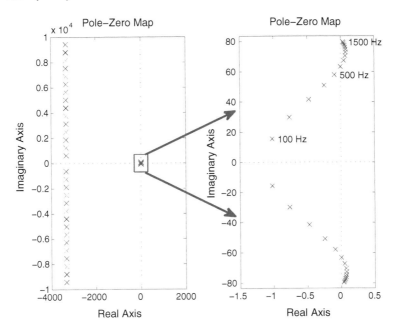

Fig. 8.2 Plot of the system poles as a function of the applied frequency ©2010 IEEE

$$
\begin{aligned}
I_{\gamma 0} &= I_{q0} \sin \delta_0 \\
I_{\delta 0} &= I_{q0} \cos \delta_0
\end{aligned}
\tag{8.7}
$$

Bearing in mind that $V_{\gamma 0} = 0$, equations that correlate operating voltage and currents in the controller's reference frame with operating speed are given as:

$$
\begin{aligned}
-\frac{R}{L} I_{\gamma 0} + \omega_0 I_{\delta 0} - \frac{\psi_m}{L} \omega_0 \sin \delta_0 &= 0 \\
-\frac{R}{L} I_{\delta 0} - \omega_0 I_{\gamma 0} - \frac{\psi_m}{L} \omega_0 \cos \delta_0 + \frac{1}{L} V_{\delta 0} &= 0
\end{aligned}
\tag{8.8}
$$

For different values of the operating speed ω_0 steady-state quantities are calculated using (8.6), (8.7) and system of Eq. (8.8). Finally, eigenvalues of the open-loop system represented by the state-space model (8.2) are plotted in Fig. 8.2 for the example of the test motor.

In the left-hand plot in Fig. 8.2 are presented system poles for different operating frequencies (100–1500 Hz). Two groups of the system poles can be distinguished. In the first group are very well damped *electrical* poles which represent fast dynamics of the motor's electrical subsystem. The right-hand plot represents a zoom-in on the other group of poles. These *mechanical* poles, which are associated with dynamics of the electromechanical subsystem, are poorly or negatively damped and have the dominant impact on motor stability.

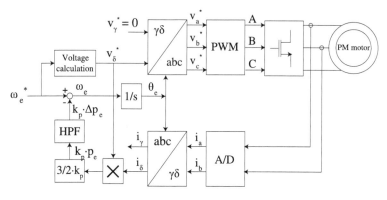

Fig. 8.3 Block diagram: modified V/f control ©2010 IEEE

8.3 Stabilization Control

In order to add the necessary damping into the system, the reference frequency is modulated using perturbations of the active power. The method has been repeatedly analyzed in literature [4, 8–13]; it will be shortly recounted here (Fig. 8.3).

After analyzing a reduced model of PM machine [4, 10] it can be observed that the damping can be achieved by modulating applied frequency proportionally to the time derivative of rotor frequency perturbations [10]:

$$\Delta\omega_e = -K\frac{d\Delta\omega_t}{dt} \tag{8.9}$$

Since measurements of the speed are not available, the desired modulating term can be acquired from perturbations of the active power p_m, thus:

$$\Delta\omega_e = -k_p\Delta p_m = -k_p\left(J\omega_0\frac{d\Delta\omega_r}{dt} + 2k_{T1}\omega_0\Delta\omega_r + k_{T0}\Delta\omega_r\right) \tag{8.10}$$

Finally, perturbations of the output power will be shown in the input, electrical power p_e, which is measurable:

$$\Delta p_m \approx \Delta p_e = \frac{3}{2}\Delta\left(v_\delta i_\delta + v_\gamma i_\gamma\right) = \frac{3}{2}V_{\delta 0}\Delta i_\delta \tag{8.11}$$

With the steady-state values calculated as in the previous section and the value of the feedback gain set as:

$$k_p = \frac{const}{\omega_0}, \tag{8.12}$$

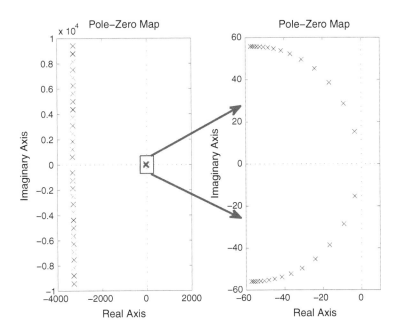

Fig. 8.4 Poles of the voltage-controlled motor including frequency modulation, $k_p = 4500/\omega_0$ ©2010 IEEE

eigenvalues of the new system are plotted in Fig. 8.4. The electrical poles do not practically change their value while mechanical poles are well-damped, having damping of about 0.7 at high frequencies.

8.4 I/f Control Method

For the given application, implementing the V/f control method was troublesome, mainly due to difficulty of setting adequate voltage references throughout the large speed range of the motor. Calculation of the voltage from required torque/current invariably depended on the values of stator resistance, which changes with temperature, and inductance, which could not be accurately measured. Additional voltage was needed for fast acceleration. During operation, overvoltage would activate overcurrent protection, undervoltage would result in motor stalling. Eventually, a lot of tuning of the reference voltage was needed to match different operating regimes and motor acceleration rate was extremely limited.

Therefore, the I/f control method has been used. Since required current depends solely on load torque and permanent magnet flux (Eq. (8.6)), which both can be accurately measured in the test setup, it was far simpler to set current as the reference instead of voltage.

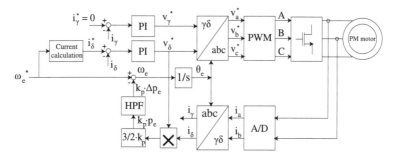

Fig. 8.5 Block diagram: I/f control with frequency modulation ©2010 IEEE

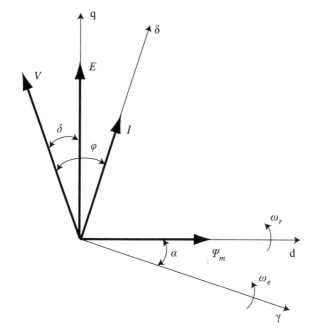

Fig. 8.6 Rotating reference frames, current control ©2010 IEEE

A block diagram of the control system is given in Fig. 8.5. The frequency modulation is retained and only current regulators were added with respect to the V/f method. Coefficients of PI regulators are adjusted so that the dynamic of the current controller loop is much faster than the dynamics of the mechanical subsystem. Value of the δ-axis reference current is set with respect to operating speed and, as a result, in steady state $\gamma\delta$ frame is connected to the current phasor (Fig. 8.6).

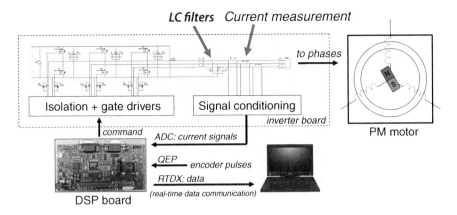

Fig. 8.7 Scheme of the test setup ©2010 IEEE

8.5 Controller Implementation and Experimental Results

8.5.1 Description of the Test Setup

Here, a short overview of the test setup will be given with focus on properties which are important for the controller implementation. Details on the test setup and experimental results are presented in Chap. 9.

The setup consists of three parts: a 2-pole slotless PM motor which shares the same housing with static air bearings, a high-frequency inverter which drives the motor and a DSP-based controller board. The motor is thoroughly described in Chap. 7 and more details on the bearings and inverter can be found in Chap. 9.

The motor was designed to support high-speed micro-milling with a maximum speed of 200.000 rpm; properties of the motor are given in Table 7.1 in Sect. 7.4. As pointed out in Sect. 7.3, it is expected that the main part of the load during milling at the maximum speed comes from the dragging torque caused by air-friction and eddy currents in the stator core. The load torque from the micro-milling itself is very small and can be regarded as a high-frequency (synchronous) disturbance (Fig. 8.7).

The gate-driving circuitry of the high-frequency inverter introduces a considerable signal delay of about 800 ns. An LC filter with the resonant frequency of 13 kHz is placed at the output of each inverter leg.

The microcontroller used is the Texas Instruments 100 MHz TMS320F2808 fixed-point DSP and its processing power has represented the main limitation for complexity of the control algorithm. Namely, at the desired switching frequency of 100 kHz this microcontroller would allow only 1000 micro-instructions; for instance, only one fixed-point sinus operation requires at least 40 micro-instructions. Therefore, stringent code implementation of the control algorithm was necessary to accommodate processing of signals within one switching period.

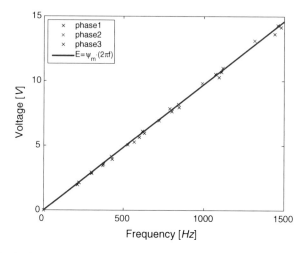

Fig. 8.8 No-load voltage—measurements ©2010 IEEE

Although speed measurements have not been used to control the motor, a reflective optical sensor was used for test-measurements of speed and its output was fed into the DSP. By using RTDX (*Real-Time Data eXchange*) communication, real-time transfer of data between the DSP and a computer has been enabled.

8.5.2 I/f Controller Implementation

Stabilization control (modulating reference frequency) was implemented as described in Sect. 8.3 with the feedback gain k_p set according to Eq. (8.12). Coefficients of PI regulators were set using the IMC strategy [14] with the goal of ensuring fast dynamic of the current controller.

For measuring permanent magnet flux and load torque decay-speed test was used. The current reference was set to a fixed value (1.5 A), the motor was driven to a very high speed and then, after switching off the inverter, frequency decay versus time and no-load voltage versus frequency was being registered. Measurements of the phase no-load voltage is given in Fig. 8.8 and decay of the rotational frequency over time in Fig. 8.9.

In order to estimate the load torque the frequency was first correlated with time using a polynomial curve-fitting of the measured data. The load torque was then calculated as:

$$T_{load} = -2\pi J \frac{df}{dt}. \tag{8.13}$$

Finally, by fitting the correlation between the load torque and frequency with a linear function—Eq. (8.6)—coefficients k_{T1} and k_{T0} are obtained (Fig. 8.10).

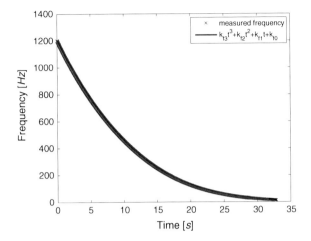

Fig. 8.9 Frequency decay—measurement ©2010 IEEE

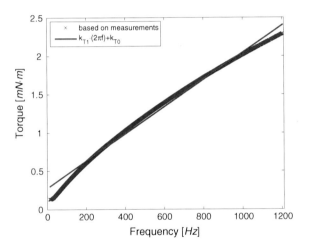

Fig. 8.10 Load torque versus frequency ©2010 IEEE

The correlation between the rotational frequency and current reference is obtained from the torque equilibrium equation:

$$I_{\delta 0} = \frac{2}{3\psi_m} (k_{T1}\omega_0 + k_{T0})\, k_{err} k_{add} \tag{8.14}$$

In (8.14) $k_{err} = 1.05$ represents a correction of the estimated torque for the maximum error of the fitting (5 %) and k_{add} is $5 \div 10\%$ added torque that is needed for acceleration and to balance some additional load torque resulting from armature-induced eddy currents.

Fig. 8.11 A phase current at
200 Hz (12000 rpm) without
and with frequency modulation ©2010 IEEE

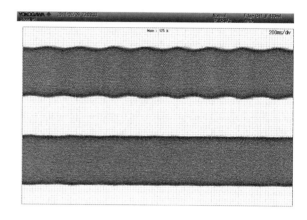

8.5.3 Experimental Results

The proposed method proved to be very convenient for control of the high-speed
motor. Only a few sets of measurements of no-load voltage and speed decay were
sufficient to estimate an adequate current reference.

A relatively large current amplitude (1 A) was used for the rotor initial positioning
and speeding up. The initial aligning of the rotor with the stator flux was done by
applying the current phasor in two directions with 90° span consecutively. The current
was reduced after speed-up and further acceleration was performed with current set
as in Eq. (8.14).

Since the current regulator takes care of adjusting the reference current to the load
torque, no compensation of the phase voltage drop was needed. At the same time,
the load torque and no-load voltage, according to which the current reference was
set, hardly changed during operation. Faster acceleration could easily be achieved
by adequately increasing coefficient k_{add} during speed-up.

Without frequency modulation high oscillations in the torque were already present
at rather low speeds (Fig. 8.11). After the oscillations would grow, the synchronism
would be lost at about 300 Hz (18.000 rpm). However, with the frequency modulation
applied and the feedback gain set as in (8.12), the torque and speed remained stable
throughout the whole speed range. The rotor speed also remained stable when it was
disturbed when touching metal objects (Fig. 8.12).

The motor was successfully drive up to highest speed achieve with this setup—
approximately 2600 Hz (156.000 rpm). There was no need for change or tuning of
initially obtained current-reference parameters and change of the stator temperature
did not affect controller performance.

This setup was inadequate to offer comprehensive tests of the controller performance such as bandwidth of the speed controller and response to changes in the load
torque.

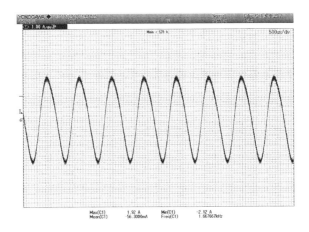

Fig. 8.12 Current waveform at 100000 rpm ©2010 IEEE

8.6 Conclusions

For very high-speed PM synchronous machines open-loop control methods are generally used so as to facilitate controller implementations in standard microcontrollers. The V/f control method with modulation of the reference frequency has proven its merits and has been regularly used for high-speed PM machines. Still, the method offers no means for controlling current (torque) while setting a suitable voltage reference with respect to the applied frequency is parameter-dependent and/or computationally difficult.

This chapter presents a new realization of the I/f method in which the voltage reference is generated through a current regulator while retaining the frequency modulation as in the V/f method. With a fast dynamic of the current control the stability of the system is preserved with the apparent advantage of having supervision of the currents incorporated. The main benefit, however, is that the required reference current can be quite simply and adequately determined using knowledge or measurements of the machine PM flux and load torque.

The method was applied in control of a small high-speed PM motor up to the rotational speed of almost 160.000 rpm. The control algorithm was easily implemented in a non-expensive, standard digital signal controller. Based on measurements of no-load voltage and speed decay the adequate current reference was set. The control was stable and good in disturbance rejection throughout the whole speed range.

References

1. P. Pillay, R. Krishnan, Application characteristics of permanent magnet synchronous and brushless dc motors for servo drives. IEEE Trans. Ind. Appl. **27**(5), 986–996 (1991)
2. A. Borisavljevic, H. Polinder, J.A. Ferreira, Realization of the I/f control method for a high-speed permanent magnet motor, in *Proceedings of the International Conference on Electrical Machines, ICEM 2010*, 2010
3. P. Mellor, M. Al-Taee, K. Binns, Open loop stability characteristics of synchronous drive incorporating high field permanent magnet motor. IEE Proc. B Electr. Power Appl. **138**(4), 175–184 (1991)
4. P. Chandana Perera, F. Blaabjerg, J. Pedersen, P. Thogersen, A sensorless, stable V/f control method for permanent-magnet synchronous motor drives, in *Applied Power Electronics Conference and Exposition, 2002. APEC 2002. Seventeenth Annual IEEE*, 2002
5. B.-H. Bae, S.-K. Sul, J.-H. Kwon, J.-S. Shin, Implementation of sensorless vector control for super-high speed pmsm of turbo-compressor, in *Industry Applications Conference, 2001. Thirty-Sixth IAS Annual Meeting. Conference Record of the 2001 IEEE*, vol. 2, pp. 1203–1209, 30 Sept 2001
6. L. Zhao, C. Ham, L. Zheng, T. Wu, K. Sundaram, J. Kapat, L. Chow, C. Siemens, A highly efficient 200,000 rpm permanent magnet motor system. IEEE Trans. Magn. **43**(6), 2528–2530 (2007)
7. S. Berto, A. Paccagnella, M. Ceschia, S. Bolognani, M. Zigliotto, Potentials and pitfalls of fpga application in inverter drives—a case study, in *Industrial Technology, 2003 IEEE International Conference on*, vol. 1, pp. 500–505, 2003
8. L. Zhao, C. Ham, Q. Han, T. Wu, L. Zheng, K. Sundaram, J. Kapat, L. Chow, Design of optimal digital controller for stable super-high-speed permanent-magnet synchronous motor. IEE Proc. Electr. Power Appl. **153**(2), 213–218 (2006)
9. J. Itoh, N. Nomura, H. Ohsawa, A comparison between V/f control and position-sensorless vector control for the permanent magnet synchronous motor, in *Power Conversion Conference, 2002. PCC Osaka 2002. Proceedings of the*, 2002
10. R. Colby, D. Novotny, An efficiency-optimizing permanent-magnet synchronous motor drive. IEEE Trans. Ind. Appl. **24**(3), 462–469 (1988)
11. R. Ancuti, I. Boldea, G.-D. Andreescu, Sensorless V/f control of high-speed surface permanent magnet synchronous motor drives with two novel stabilising loops for fast dynamics and robustness. IET Electr. Power Appl. **4**(3), 149–157 (2010)
12. L. Xu, C. Wang, Implementation and experimental investigation of sensorless control schemes for pmsm in super-high variable speed operation, in *Industry Applications Conference, 1998. Thirty-Third IAS Annual Meeting. The 1998 IEEE*, vol. 1, pp. 483–489, Oct 1998
13. T. Halkosaari, Optimal U/f-control of high speed permanent magnet motors, in *Industrial Electronics, 2006 IEEE International Symposium on*, vol. 3, pp. 2303–2308, 2006
14. C.E. Garcia, M. Morari, Internal model control: a unifying review and some new results. Ind. Eng. Chem. Process Des. Develop. **21**(2), 308–323 (1982)

Chapter 9
Experimental Results

9.1 Introduction

Experiments and accurate measurements of high-speed electrical machines often meet many practical difficulties mainly as a consequence of constrained geometries and the fast rotation itself. To test many of the phenomena discussed in this thesis, dedicated test setups would be necessary although that would exceed the scope of the project. The setup that was developed for the purpose of the thesis project was primarily designed to demonstrate the new spindle-drive concept and test the overall motor performance. With the experiments reported in this chapter, the test capabilities of the designed setup were explored to gather practical data on the motor and bearings, which would further allow an assessment of the developed models and design approach.

The chapter starts with description of the practical setup including a short description of stator and rotor assembly and important information about the air-bearing setup and high-frequency inverter. In Sect. 9.3 measurements of the motor phase impedance, i.e. phase resistance and inductance, are reported.

The main tests performed for verification of the models developed in this thesis are the speed-decay and locked-rotor tests as reported in Sects. 9.4 and 9.5. The speed-decay tests were performed to measure the no-load voltage and losses of the motor. The locked-rotor tests were used to measure apparent phase impedance at high frequencies and to assess motor losses under load.

Finally, the overall performance of the electric drive during rotation is reported and discussed.

A. Borisavljević, *Limits, Modeling and Design of High-Speed Permanent Magnet Machines*, 175
Springer Theses, DOI: 10.1007/978-3-642-33457-3_9,
© Springer-Verlag Berlin Heidelberg 2013

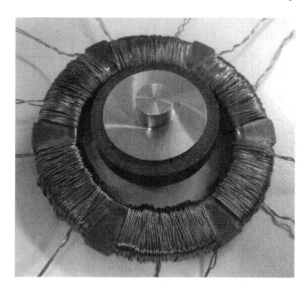

Fig. 9.1 A photo of the stator and rotor; the rotor is without the sleeve ©2009 IEEE

9.2 Practical Setup

9.2.1 Stator Assembly

The stator cores are made of 0.1 mm laminations of non-oriented *Si*-steel 10JNEX900 [1]. The lamination sheets were cut to the design dimensions (see Fig. 7.11) and glued together (Figs. 9.1 and 9.2).

The stator cores are wound toroidally using two parallel self-bonding wires Thermibond® 158 [2]. A smaller number of turns per coil than initially designed (40 instead of 44) was applied due to a restriction on the stator's total axial length imposed by the bearing housing.

Windings were finally pressed and cured by means of a DC-current surge which was adjusted to heat the wires to 230 °C for approximately 3 min. The high temperature causes softening of the adhesive varnish before curing and, in turn, eliminates the need for impregnation of the windings.

9.2.2 Rotor Assembly

The iron shaft of the rotor was first produced according to the dimensions from Fig. 7.10. The PPS-bonded NdFeB magnet was applied onto the incised part of the rotor disc using the injection-molding technique.

Fig. 9.2 A photo of the final rotor

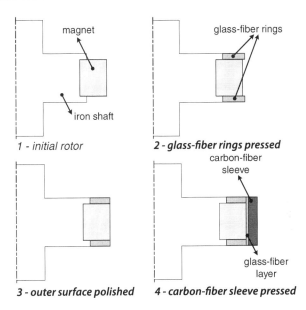

Fig. 9.3 Steps in the assembly of the rotor retaining sleeve

The realization of the rotor bandage was the most demanding part of the rotor production. Section 7.6 includes an explanation of the procedure which was eventually used to ensure the structural integrity of both the plastic-bonded magnet and carbon fibers; the procedure is depicted in Fig. 9.3. On the other hand, the pressing of the sleeve parts during assembly was expected to cause a large mass unbalance in the rotor.

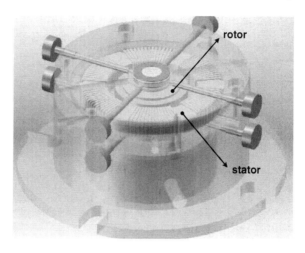

Fig. 9.4 The motor inside the air-bearing housing (taken from 7.2)

For test purposes one rotor was produced in which the magnet was substituted with epoxy (*dummy* rotor) and this rotor was used for measuring air-friction losses and assessing losses in the magnet.

9.2.3 Air-Bearings Test Setup

A setup with air bearings for testing the motor was developed as a master project assignment [3]. The concept of the setup was presented in Sect. 7.2.

The air-bearing housing was designed to accommodate the stator and facilitate good thermal contact with environment. A transparent image of the air-bearing housing with the motor in it is given in Fig. 9.4 and a photo of the setup is shown in Fig. 9.5.

Two important physical requirements were set for the air bearings: the bearings were supposed to provide a sufficient load capacity to support the PM rotor and to ensure stability of rotation for a maximum possible range of rotational speeds.

While the stiffness of air bearings is generally high, their load capacity is rather low and this holds particularly true for air bearings with simple orifice restrictors, the type employed in this test setup. To define maximum radial load force on the rotor, the expression (3.81) for unbalanced magnetic force in the motor was used. With $\Delta x_{max} = 0.5\,\mathrm{mm}$ of maximum anticipated rotor eccentricity (1 mm is the physical maximum) the highest possible value of the unbalanced force was estimated (Table 9.1). Finally, the air bearings were designed so their load capacity is, at least, an order of magnitude higher than the estimated value.

The main concern with the bearings was their influence on the stability of rotation. Instability problems with journal fluid bearings were discussed in Sect. 6.3 and a simplified theoretical explanation for the phenomena was given in Sect. 5.3.2: because

Fig. 9.5 A photo of the air-bearing setup ©2009 IEEE

Table 9.1 Estimations of dynamic parameters of the motor and the bearings

Parameter	Analytical	FEM
Maximum unbalanced force (N)	0.5780	0.3638
Radial load capacity (total) (N)[a]	8.2	8.4
Radial stiffness (total) (kN/m)[a]	1170	1200
Critical frequency (Hz)	878.5	889.7

[a] Values obtained from [3]

of non-synchronous aerodynamic forces in the bearings, rotors in aerodynamic- and lubricated journal bearings tend to become unstable at speeds very close to twice their critical speed. On the other hand, static air bearings with orifice restrictors have not yet been associated with any type of instability [4]. Nevertheless, a few reported high-speed spindles supported by static air bearings were intentionally designed to maintain their rotation below twice the first critical speed [5, 6]: the authors expected that aerodynamic forces would overpower static bearing forces at very high speeds.

The critical frequency of the test rotor is estimated by:

$$f_{cr} = \frac{1}{2\pi}\sqrt{\frac{2k_1}{m}} = \frac{1}{2\pi}\sqrt{\frac{k}{m}}, \tag{9.1}$$

where m is the rotor mass, k_1 is the stiffness of a single radial bearing and $k = 2\,k_1$ is the total radial stiffness of the air bearings.

Based on the models developed in [3] the first critical frequency is estimated to be somewhat below 900 Hz (54.000 rpm; Table 9.1) and potential stability problems were expected in the speed range of 110.000–120.000 rpm.

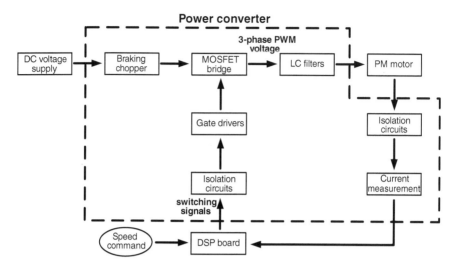

Fig. 9.6 A functional scheme of the developed inverter [3]

9.2.4 High-Frequency Inverter

At the time of the motor design there was no power converter on the market that would be suitable for driving such a high-speed synchronous PM motor. Therefore, a high-frequency PWM inverter was developed within the project as a master project assignment [3]. The inverter was also supposed to be capable of driving the high-speed motor from the AMB setup described in [7]. Moreover, an in-house built inverter provided the possibility to develop a dedicated controller for the spindle motor (Fig. 9.6).

A functional scheme of the inverter is shown in Fig. 8.7. The core of the design is a MOSFET bridge capable of 200 kHz switching and tested for DC voltages up to 300 V. The output phase currents of the inverter are filtered by LC filters and current measurements are provided by the developed high-bandwidth circuitry.

9.3 Motor Phase Resistance and Inductance

9.3.1 Phase Resistance

Measurements of the motor phase resistance were done using a simple setup shown in Fig. 9.7. The current in the series connection of two phases was adjusted using a voltage source while voltage and current were also measured with an oscilloscope and current probe, respectively. Higher currents resulted in higher winding temperature

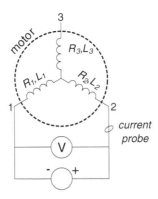

Fig. 9.7 Phase-resistance measurement setup

which was monitored using a thermocouple. The measurements were repeated for all three pairs of the phases.

Motor resistance can also be estimated using the equations presented in Sect. 3.6.2 on motor copper losses. Resistance of a single phase is given by (see Eq. (3.110))[1]:

$$R = \frac{4}{nd_{Cu}^2\pi} l_{Cu}\rho_{Cu},\tag{9.2}$$

where the length of a single phase conductor can be estimated as:

$$l_{Cu} = [2l_s + (r_{so} - r_s)\pi]\,2\,N.\tag{9.3}$$

Copper resistivity changes linearly with temperature:

$$\rho_{Cu} = \rho_0\,(1 + \alpha_{Cu}\Delta T)\tag{9.4}$$

and for a base temperature of 20 °C ($\Delta T = T - 20\,°C$) the thermal coefficient of the resistivity is $\alpha_{Cu} = 4.03 \cdot 10^{-3}\ \Omega m/°C$.

Predicted and measured values of the series resistance of two motor phases with respect to temperature are presented in Fig. 9.8. The resistance values stemming from the model fall well within the range of the measured values. Apparently, Eq. (9.3) is quite adequate for the estimation of conductor length.

From the measurements shown in Fig. 9.8 there is a noticeable difference in the resistances of different phases. This can certainly be ascribed to the manual winding process; apparently there is a slight variation in the number of turns of toroidal coils in the stator.

---

[1] Notation from the previous chapters is maintained.

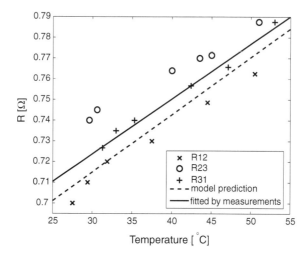

Fig. 9.8 Measured and predicted values of phase-to-phase motor resistance

The measured values were fitted with a linear curve to represent the actual phase resistance of the motor. The measurement-fitted resistance yielded 0.3486 Ω at 20 °C and the predicted value for the same temperature is 0.3436 Ω.

9.3.2 Phase Inductance

An accurate measurement of a motor DC inductance is difficult to achieve since it would require measuring the winding flux linkage. Assessment of the inductance was attempted using the following three measuring methods.

Firstly, the inductance between two motor phases was measured using an RLC meter. This measurement seemed rather inaccurate: the meter showed largely different results in different measurement modes (series or parallel inductance) therefore this measurements were abandoned.

This measurement indicated, however, that the setup suffers from very large eddy-current losses in the housing. Because of the toroidally-wound stator fitted into a highly conductive, aluminum housing, armature leakage flux causes rather strong eddy currents in the housing. The difference in RLC-meter inductance readings on the machine with and without the top housing was tremendous.

Another method to measure inductance was to include the motor windings in a resonant circuit; the measurement setup is shown in Fig. 9.9. The frequency of the AC voltage source was increased until a maximum amplitude of the current (minimum impedance) was registered. Given the resonant frequency of the circuit f_{res}, the inductance of a single motor phase was estimated by:

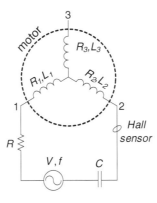

Fig. 9.9 Phase-inductance measurement setup: resonant circuit

Table 9.2 Phase-inductance values measured with the resonant circuit and the predicted value

Capacitor (μF)	Resonant frequency (Hz)	Inductance—measured (μH)	Inductance—predicted (μH)
30	1660	153	43
60	1240	137	

$$L = \frac{1}{2 (2\pi f_{res})^2 C}. \tag{9.5}$$

Phase inductance estimations for two values of the capacitor are given in Table 9.2. Large deviations of the estimated values from the model prediction are evident.

Finally, inductance was estimated using a motor short-circuit test. Namely, the windings of the machine were short-circuited, the rotor was levitated in the air bearings and rotated using an air-blow gun. The frequency and amplitude of the current were registered and the phase inductance was calculated as:

$$L = \frac{1}{\omega} \sqrt{\left(\frac{\hat{e}}{\hat{i}}\right)^2 - R^2} = \frac{1}{\omega} \sqrt{\left(\frac{\omega \psi_{max}}{\hat{i}}\right)^2 - R^2}. \tag{9.6}$$

The measurement of the phase resistance R were presented in the previous subsection; the flux-linkage amplitude ψ_{max} was also measured; the results are shown in the following section.

This measurement was rather inaccurate since it was very difficult to maintain rotor speed using the air gun. The maximum achievable rotational speed was quite low (≈ 38 Hz) which, in turn, resulted in a very small phase reactance and additionally decreased the measurement accuracy. Estimations of the phase inductance based on this test varied greatly, between 150 and 350 μH.

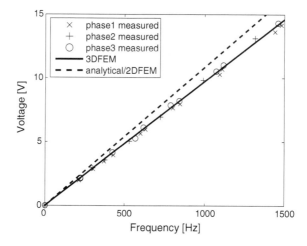

Fig. 9.10 No-load phase voltage: measurements and modeling predictions

In conclusion, the first and last method appeared to be rather inaccurate; the results obtained by the resonant-circuit method will be maintained by the locked-rotor measurements in Sect. 9.5, however, those results are valid only at the frequencies corresponding to the resonant frequencies of the established circuits.

9.4 Speed-Decay Tests

Speed-decay tests were performed to measure the no-load voltage and losses of the motor. These tests were also used to define the current reference for the controller as described in Chap. 8.

Hence, in a speed-decay test the rotor is driven to a certain speed, the drive (inverter) is then switched off and the rotational-frequency decay versus time and no-load voltage versus frequency is registered.

Measurements of the no-load phase voltage are shown in Fig. 9.10. The measured voltage matches the prediction of the 3D FE simulations; analytical and 2D FE model overestimated the motor voltage due to axial flux leakage (see Sect. 7.5). Such a good match between the measured and simulated motor voltage can be ascribed to very accurate data on magnet BH curve which was measured after the magnet application on the rotor.

For measurements of no-load losses, the decay of rotational frequency in time was registered. The frequency was correlated with time using a polynomial curve-fitting of the measured data (see Fig. 8.9 in Chap. 8):

$$f(t) = k_{f3}t^3 + k_{f2}t^2 + k_{f1}t + k_{f0}. \tag{9.7}$$

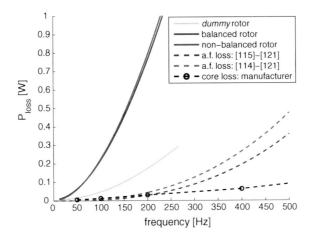

Fig. 9.11 No-load losses of the motor at low speeds based on the speed-decay tests; *full lines* represent losses based on measurements and *dashed lines* represent predictions

The drag torque was then calculated by:

$$T_{drag} = -2\pi J \frac{df}{dt} \tag{9.8}$$

and the power of the drag (= no-load losses) was found as:

$$P_{drag} = 2\pi f T_{drag}. \tag{9.9}$$

Firstly, the rotor without the magnet (*dummy* rotor) was tested to assess air-friction loss. It was spun by the air gun up to maximally ~280 Hz (~16.800 rpm) and the speed-decay was registered. The loss power, which consists of air-friction loss only, was calculated and compared with the analytical predictions of models presented in Sect. 3.6.3, Fig. 9.11. In the figure no-load losses are also shown for the complete rotor in the same speed range ($0 \div 500$ Hz).

It is evident from Fig. 9.10 that models of air-friction losses based on empirically developed formulas (Sect. 3.6.3) cannot account for the air-friction losses in the test motor. However, those models were developed for rotating conditions that significantly differ from the conditions in the test machine: the test rotor is supported by air bearings in all directions, influencing greatly the air flow.

From the measured no-load losses it is evident that electromagnetic losses dominate the low frequency operation, as expected. The losses in the machine with a non-balanced rotor were slightly higher than with a rotor which is mechanically balanced: some energy is lost in the bearing dampers during rotor synchronous whirling [8].

In Fig. 9.12 no-load losses are plotted for frequencies up to 1200 Hz (72.000 rpm). In this figure, measured air-friction losses, using the *dummy* rotor, are extrapolated to account for the loss increase at higher frequencies. It is hardly possible to obtain

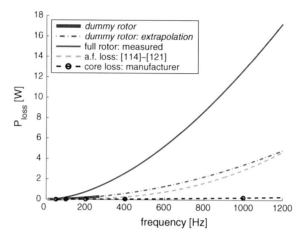

Fig. 9.12 No-load losses of the motor based on the speed-decay tests up to 1200 Hz (72.000 rpm); *full lines* represent losses based on measurements, the *dashed-dotted line* represents measurement-based extrapolation and *dashed lines* represent predictions

an accurate prediction using such an extrapolation: at lower speeds, where these losses are measured, the air flow in the air gap is predominantly laminar and the loss increase is closely quadratic to the speed; this would surely not be the case when strong turbulences are present. Still, it is quite certain, judging from both Figs. 9.11 and 9.12, that overall no-load losses in the setup cannot be explained merely as a combination of air-friction- and stator-core losses.

Eddy-current losses in the conductors could not be directly measured, but those losses hardly represent a large portion of the overall losses: both analytical and FEM models (see Sect. 7.5) suggest extremely low eddy-current losses in the optimized conductors (below 0.1 W for the frequencies shown in Fig. 9.12); these predictions are, for that reason, not shown in the figures.

To account for unexpectedly high no-load losses, transient 3D FE simulations were performed, this time taking into account also the presence of the housing. Namely, top and bottom aluminum housings of the air bearings are placed only 0.5 mm above and below the axial sides of the rotor disc, see Fig. 9.13. Since the axial leakage of the permanent-magnet flux is, apparently, rather pronounced in this motor, it is reasonable to expect significant losses in the housing in regions close to the magnet.

Losses in the housing were calculated using a transient 3D FEM and the results for the low frequency range $(0 \div 600 \text{ Hz})$ are plotted in Fig. 9.14 along with measurements and estimations of other dominant losses. It is difficult to judge the combined contribution of the loss factors since the information about the various losses is obtained in different ways: measurements, analytical and FE modeling and manufacturer's data. Still, it can be inferred from the last plot that air friction and eddy currents in the housing are predominant sources of no-load losses. Unfortunately, both of those loss factors could not have been analytically modeled: housing losses

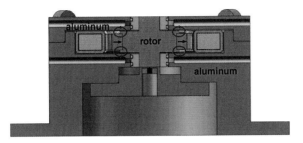

Fig. 9.13 Cross-section of the test setup; field lines of the permanent-magnet field are sketched at regions where housing losses are expected

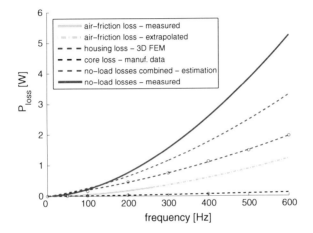

Fig. 9.14 No-load losses of the motor at low speeds including 3D FEM estimation of losses in the housing; *full lines* represent losses based on measurements, the *dashed-dotted line* represents measurement-based extrapolation and *dashed lines* represent predictions

are essentially a 3D phenomenon and no adequate model is available for the air friction in the test motor.

Causes for discrepancy between the measured no-load losses obtained by the speed-decay tests and the estimation of the combined losses (obtained from the measurements and extrapolation of the air-friction loss and the estimations of housing and stator-core losses) can be sought in various possible factors. It is quite conceivable that the manufacturer's data underestimate losses in the core in a certain extent: the stator manufacturing process has an influence on the core properties; rotation of the field in the core additionally increases the losses (the manufacturer's data are, namely, given for an unidirectional magnetic field). Extrapolation of the air-friction measurements is not a reliable means for the loss assessment as it was already emphasized: the air-friction loss rises more rapidly at high frequencies than the low-frequency trend of the increase suggests. Finally, the transient 3D FEM for the

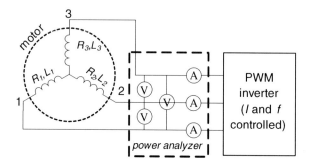

Fig. 9.15 Locked-rotor test: measurement setup

housing-loss calculation used rather simplified geometry—an intricate model would, however, require too much time and computation resources.

9.5 Locked-Rotor Tests

To assess losses in the motor under load, locked-rotor tests were performed. A scheme of the test setup is shown in Fig. 9.15. The inverter described in Sect. 9.2.4 was used to set currents of different frequencies in the motor windings. Active power P and reactive power Q flowing into the motor were measured using a power analyzer. The temperature in the windings was also monitored using a thermocouple. From these measurements apparent (AC) phase resistance and inductance were determined as:

$$R_{AC} = \frac{P}{3 I^2},\tag{9.10}$$

$$L_{AC} = \frac{Q}{3 (2\pi f) I^2}.\tag{9.11}$$

The AC resistance represents load-dependent losses in the test setup at the given current and electrical frequency and the AC inductance tells about the influence of eddy-currents on the armature field of the motor since the field of the eddy-currents counteracts the stator-current field and, in effect, reduces apparent phase inductance.

The test results for frequencies up to 3500 Hz and rms currents up to 2.5 A are shown in Figs. 9.16 and 9.17. The AC resistance is normalized over the DC resistance which has been corrected for the temperature impact using measurements of the winding temperature:

$$r_{AC} = \frac{R_{AC}}{R_{DC}} = \frac{R_{AC}}{R_0 (1 + \alpha_{Cu} \Delta T)},\tag{9.12}$$

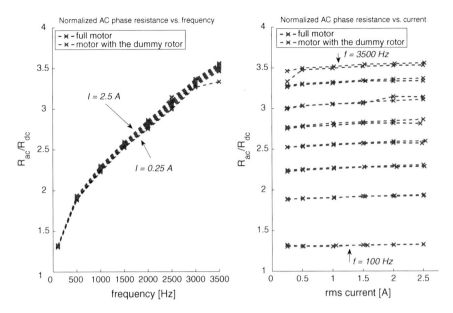

Fig. 9.16 Normalized AC resistance of the motor measured in the locked-rotor test

where the DC phase resistance at room temperature R_0 was measured (Sect. 9.3) and the AC resistance R_{AC} was calculated from the measurements according to Eq. (9.10).

The tests were initially performed on the original setup—the motor fitted into the air-bearing housing—and they were repeated for two different rotors: the normal, PM rotor and the rotor in which the magnet has been replaced by epoxy (*dummy rotor*).

Several conclusions can be made based on the measurements. Firstly, it is quite evident that the presence of the magnet in the rotor does not influence the results. Hence, neglecting losses in the plastic-bonded magnet done in the modeling is thoroughly supported by the measurements. Moreover, taking into account the very low conductivity of carbon-fiber composites and the FE calculations from Sect. 7.5 (showing rotor-iron losses be negligible), it is quite suitable to disregard rotor losses in the designed machine.

On the other hand, the measurements show that eddy-currents are present in the setup to a great extent. The influence of eddy-currents is clearly visible in the measurements of the phase inductance which decreases greatly with frequency. At the targeted frequency of 3300 Hz (corresponding to 200.000 rpm) the overall frequency-dependent losses amounted to almost 3.5 times the value of the classical conduction loss ($I^2 R_{DC}$) in the machine.

Both the measured apparent resistance and inductance changed very little with current (except the inductance values at very low frequencies). The small deviations

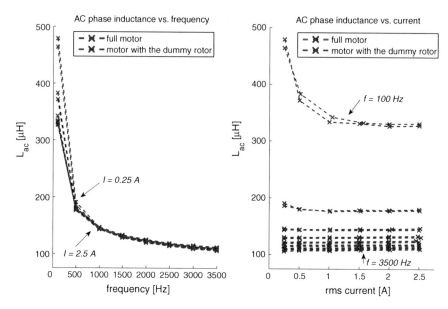

Fig. 9.17 AC inductance of the motor measured in the locked-rotor test

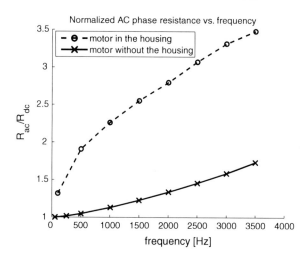

Fig. 9.18 Normalized AC resistance of the motor measured in the locked-rotor test: with and without housing

were influenced not only by the armature field but also by the temperature in the core and the housing during different measuring sessions.

Already during measurements of the phase inductance (Sect. 9.3) it was noticed that the presence of the housing has an impact on the armature field. Therefore, the locked-rotor test was repeated for the motor outside the housing. The averaged

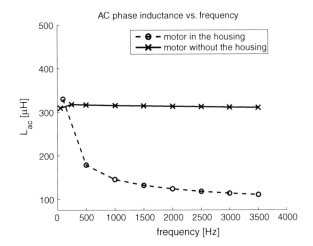

Fig. 9.19 AC inductance of the motor measured in the locked-rotor test: with and without housing

measured values at different frequencies from two test setups are shown in Figs. 9.18 and 9.19.

By comparing the two test cases it is evident that the bulk of eddy-current losses is generated in the housing: without the housing the change of the AC phase inductance with frequency is insignificant. The frequency-dependent portion of losses under load at the maximum motor frequency (speed) would add less than 70 % to the regular DC conduction loss for the motor outside the housing; with the housing, however, this figure amounts to 250 %.

Armature field is generally considered to have a small impact on the performance of slotless machines; this seems, however, not quite correct for toroidally-wound machines—at least for those placed in an electrically-conductive housing. A magnetostatic model is adequate to describe performance of the test motor on itself, but cannot account for losses that are induced outside the machine, losses which appear to be very high in this case.

Measurements of the phase inductance on the motor without housing agree with the 3D FEM predictions (Table 7.3, Sect. 7.5). The total DC inductance was estimated from the measurements to be around 325 μH and is just several percent lower than the 3D FEM value— 350 μH.

The measured increase of the apparent phase resistance in the motor without housing still cannot be explained using analytical and FE models so far developed in the thesis. All the models agree that the armature field in the stator iron is very small, the field at which losses in the iron should be negligible. On the other hand, it is not reasonable to guess that the proximity effect has such a significant influence on the AC copper losses given the small influence of the armature field in the air gap.

It was observed in Sect. 9.3 that measurements of the resistances of different phases indicate some unbalance between the phases: the toroidal coils appear to be

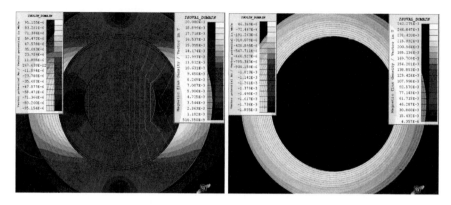

Fig. 9.20 Flux-density plots and field lines of the motor with balanced coils and with a single coil having an additional turn—2D FEM

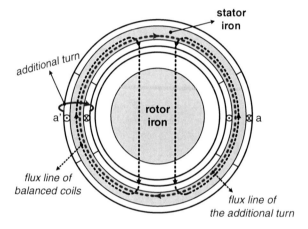

Fig. 9.21 Motor geometry cross-section: depiction of the effect of an additional turn in one of the phase coils on the field in the stator core

unevenly wound as a result of the manual production. In order to assess the influence of unbalanced coils in the motor, the 2D FE model from Sect. 7.5.1 was adjusted to account for small variations of numbers of turns of the coils.

In Fig. 9.20 a flux-density plot and field lines are shown for examples of an evenly wound machine and of a machine with only one extra turn in one of the two toroidal coils of the phase a (precisely, 41 turns instead of 40). While adding an extra turn in the left-hand coil of the phase does not increase by much the phase inductance (76–64 μH, as calculated by the 2D FEM), flux density in the stator iron increases tremendously: 262 mT of the maximum flux-density in the unevenly wound motor against only 20 mT in the balanced machine (simulated for $i_a = 1$ A, $i_b = i_c = -0.5$ A). Finally, the losses in the core will quadratically follow the increase in

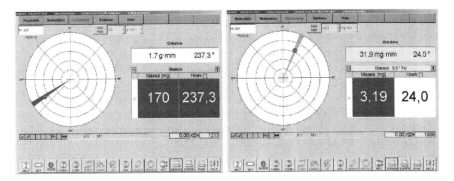

Fig. 9.22 Measured rotor static unbalance before and after mechanical balancing

amplitude of the flux density which can, therefore, explain much higher iron loss than expected.

The effect is also depicted in Fig. 9.21: instead of exiting the stator and crossing the air gap, the field of the added turn links entirely through the stator iron whose extremely small reluctance makes this field very high. At the same time, this *additonal* field penetrates two serially-connected coils in opposite directions with respect to their orientations causing the two flux linkages of the coils to cancel each other to a great extent and, accordingly, greatly reduce the influence of the field increase on the phase inductance (Fig. 9.22).

This phenomenon implies a potential disadvantage of toroidal windings. Namely, for an efficient machine, the toroidal coils in the stator must have an equal number of turns and the winding production process should be highly repetitive.

9.6 Motor Operation and Performance

Originally, the manufactured rotor was not capable of reaching very high speeds because of rotor unbalance—the maximum rotational speed attained with such a rotor was, approximately, 330 Hz (20.000 rpm). The rotor manufacturing, which included three press-fittings of the retainment rings, resulted in a rather large unbalance: the measured static unbalance of the rotor was 1.7 g·mm which corresponds to a 45 μm shift of the inertia axis with respect to the rotor geometrical center. For the air bearings with only 14 μm clearance this was far beyond permissible. After the balancing process, however, the static unbalance was reduced to only 32 mg·mm.

The motor controller presented in Chap. 8 which used the developed realization of the I/f control method performed very well. Once set coefficients for the controller current reference were maintained in all the tests throughout the speed range. Figure 9.23 shows the setting of the current versus rotational speed. Anticipated required current at the maximum rotational speed—2.38 A—was somewhat higher than the originally predicted value of 2 A, mostly due to the 10 % reduction of the

Fig. 9.23 Phase current reference (rms) with respect to the rotational speed setting

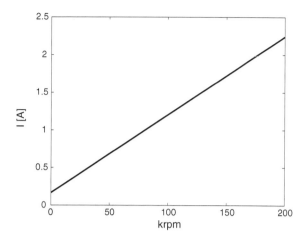

motor no-load voltage as a result of axial flux leakage (see Table 7.3). The output electromagnetic power of the rotor required to sustain the rotation at 200.000 rpm is, based on the current reference and no-load voltage, estimated at 216 W.

Initially, the critical speed was passed unnoticed since the rotor was very well-balanced. However, after rotational frequency was gradually increased beyond 2000 Hz (120.000 rpm), the rotor became noticeably unbalanced again: apparently, the rotor parts rested in a new position. The critical speed was observed in the vicinity of 850 Hz (51.000 rpm), slightly lower than predicted (see Table 9.1).

Eventually, the newly established rotor unbalance became so large that it was more and more difficult to cross the critical speed. The rotor was touching the bearings when operating near the critical speed and a higher motor torque was becoming necessary to overcome this.

The maximum speed attained with this setup was 2600 Hz (156.000 rpm). At that speed the rotor tangential speed was 270 m/s and, to the author's knowledge, that is more than previously reported tangential speeds of electrical machines. The machine operated stably up to this speed: rotor remained unheated, maximum registered temperature in the stator was 44 °C. The rotor structure seemed undamaged and the no-load voltage of the machine did not change. The designed rotor retainment fulfilled the task of enabling very high-speed rotation; on the other hand, it caused problems with the unbalance which appeared to be too high for the air bearings.

In one of the subsequent tests a major accident happened: with the rotor operating in the vicinity of critical speed, the rotor crashed into the bottom air bearing. This damaged the bearing, disturbed the alignment between the top and bottom bearings and, consequently, prohibited further testing. The last measurement data from the setup were recorded at the speed of 100.000 rpm (shown in Chap. 8).

One of the important results of this study is that the rotor remained stable at a much higher speed than twice the critical speed. It shows, thus, that static air bearings represent a good solution for stable high-speed operation and that the limits of rotational stability of such bearings are much higher than those of aerodynamic and lubricated bearings. At the same time, very tight tolerances of air bearings impose high precision standards on the rotor manufacturing.

9.7 Conclusions

The chapter reports results of practical evaluation of the test motor and the setup. Practical data from the measurements are used for assessment of the developed models and design approach.

The chapter includes a short description of stator and rotor assembly and important information about the air-bearing setup and high frequency inverter.

Measurements of the DC phase resistance confirmed the model prediction thereby verifying the model of the DC-conduction loss in the windings. Relatively large deviations of measured resistances of different phases are noticed as a result of an uneven number of turns due to errors in the manual winding process.

Direct measurements of the phase inductance were performed in different ways; however, only the method of including motor phases in a resonant circuit gave reasonably accurate values.

The main tests carried out for verification of the models developed in this thesis are the speed-decay and locked-rotor tests. Speed-decay tests were performed to measure the no-load voltage and losses of the motor.

The measured values of no-load voltage of 3D FEM matched the predicted values of no-load voltage very closely: this can also be ascribed to very accurate data on the magnet remanent field and recoil permeability.

According to the results on the drag torque obtained from the speed-decay tests with a rotor without permanent magnet, it is evident that the models generally used for more conventional machines with slender rotors fail to account for the air-friction loss in the test machine. However, those models were developed for rotating conditions that significantly differ from the conditions in the test machine.

Speed decay tests showed a great discrepancy between calculated and predicted no-load losses. With help of 3D FEM, the high no-load losses can be explained by, apparently, very pronounced losses in the housing caused by the significant axial flux leakage of the PM field. Air friction and eddy currents in the housing are predominant sources of no-load losses; unfortunately, both of those loss factors could not have been analytically modeled.

Locked-rotor tests were performed to assess the losses in the motor under load. It is easy to infer from the measurements that the presence of the magnet in the rotor does not influence either field or losses in the motor. The plastic-bonded magnet proved to be extremely resistive to eddy-current losses, which is a great advantage for high-speed applications.

Nevertheless, the measurements show that eddy-currents are present in the setup to a great extent. By comparing the losses generated in the setup with and without housing it is evident that the bulk of eddy-current losses is generated in the housing.

The losses in the aluminum housing are, actually, the weakest point of the design; along with air friction, eddy currents in the housing represent the most dominant loss factor in the setup, greatly influencing drive performance and temperature.

These measurements also showed that the magnetostatic model cannot account for all important phenomena in a toroidally-wound machine. Namely, such a model cannot adequately represent the armature field in the machine, particularly if the machine is placed in an electrically-conductive housing.

Measurements of the phase inductance on the motor without housing agree with the 3D FEM predictions.

Unevenly wound coils in motor phases caused an unnecessary increase of the armature field and losses in the stator core. This phenomenon implies that an automatic and highly repetitive winding process is strongly recommended for toroidally-wound machines.

After rotor balancing the motor was capable of reaching very high speeds. However, after rotational frequency was gradually increased beyond 2000 Hz, the rotor became noticeably unbalanced again. The problem of the recurring unbalance eventually resulted in the bearing crash during rotation in vicinity of the critical speed.

The maximum rotational speed obtained with the setup was 156.000 rpm. At that speed, the rotor developed the tip tangential speed of 270 m/s which is the highest tangential speed of an electrical machine that has yet been reported in academic literature. Very high attained tangential speed confirms validity of the approach for retaining sleeve optimization.

The motor I/f controller presented in Chap. 8 performed very well. Once set coefficients for the controller current reference were maintained in all the tests throughout the speed range.

The test setup demonstrates the suitability of aerostatic bearings for very-high-speed operation: a rotor in aerostatic bearings can operate well above twice the critical speed. This renders limits of their rotational stability higher than the limits of aerodynamic and lubricated journal bearings.

References

1. JFE Super Cores Magnetic Property Curves, JFE Steel Corporation (2003), http://www.jfe-steel.co.jp/en/products/list.html#Electrical-Steels
2. Enamelled wires: Thermibond® 158, Von Roll Isola (2004), http://products.vonroll.com/web/download.cfm?prd_id=1485&are_id=2&lng_id=EN
3. P. Tsigkourakos, *Design of a Miniature High-Speed Spindle Test Setup Including Power Electronic Converter*. Master's Thesis, Delft University of Technology, 2008
4. A. van Beek, *Machine Lifetime Performance and Reliability*. Delft University of Technology, 2004
5. D. Huo, K. Cheng, F. Wardle, Design of a five-axis ultra-precision micro-milling machine—UltraMill. Part 1: holistic design approach, design considerations and specifications. Int. J. Adv. Manuf. Technol. **47**(9–12), 867–877 (2009)

6. I. Pickup, D. Tipping, D. Hesmondhalgh, B. Al Zahawi, A 250,000 rpm drilling spindle using a permanent magnet motor, in *Proceedings of International Conference on Electrical Machines—ICEM'96*, pp. 337–342, 1996
7. M. Kimman, H. Langen, R.M. Schmidt, A miniature milling spindle with active magnetic bearings. Mechatronics **20**(2), 224–235 (2010)
8. G. Genta, *Dynamics of Rotating Systems* (Springer, Berlin, 2005)

Chapter 10
Conclusions and Recommendations

10.1 Models Presented in the Thesis

Various analytical models are presented in the thesis. In this section the models will
be reviewed; particular attention is given to model verification.

Electromagnetic Models

- In principle, magnetostatic modeling is very well-suited for representing field
  within a slotless PM machine, which is also confirmed in the thesis. The influence of the eddy currents induced in the rotor magnet and back iron as a result
  of armature-field fluctuations is negligible, particularly for a rotor with a plastic-
  bonded magnet. This is demonstrated by FE calculations in Sect. 7.5 and measurements in Sect. 9.5. On the other hand, it is shown in Chap. 9 that eddy-current
  losses in the housing have a great influence on the motor performance—both on the
  armature field and overall losses. This effect comes into play in toroidally-wound
  machines placed in an electrically-conductive housing. Since good thermal conductors used for cooling jackets and housings are also, as a rule, good electrical
  conductors, the phenomenon of losses in the housing can be expected in most
  toroidally-wound machines. Hence, the magnetostatic model cannot account for
  all important phenomena in a toroidally-wound machine and some improvement
  is needed for a fully adequate model.
- 2D modeling in a plane perpendicular to the axis of rotation is usually sufficient
  for representation of rotating machines. However, axial flux leakage of the field
  of the permanent magnet is very pronounced in the designed motor, mostly due
  to the presence of the small extrusions of the rotor iron shaft made to protect the
  magnet on its axial sides. The leakage effect causes a 10 % reduction of the motor
  no-load voltage with respect to the 2D-model (analytical and FEM) predictions.

A. Borisavljević, *Limits, Modeling and Design of High-Speed Permanent Magnet Machines*, 199
Springer Theses, DOI: 10.1007/978-3-642-33457-3_10,
© Springer-Verlag Berlin Heidelberg 2013

- The model of the PM field is confirmed by magnetostatic 2D FEM simulations and the field reduction influenced by the axial flux leakage is observed in 3D FE simulations. The predictions of 3D FEM match the measured values of no-load voltage very closely: this can also be ascribed to very accurate data on the magnet remanent field and recoil permeability.

- The armature field model developed in the thesis is not fully suitable to represent a motor with toroidal windings; as both FEM and measurements suggested, the model failed to represent certain phenomena that have a strong impact on the test-motor performance. Firstly, the model does not take into account the actual, toroidal distribution of the conductors since it neglects the magnetic field outside the stator iron. However, as seen in the FE simulations and confirmed by the measurements, external leakage of the armature field is immense, causing the motor phase inductance to be an order of magnitude higher than analytically predicted. Secondly, as already mentioned, a magnetostatic model is not suitable to represent a toroidally-wound machine in an electrically-conductive housing.

- The analytical model of the distorted flux density in a slotless PM machine with an eccentric rotor, based on the simple approximation of the relative permeance function (Sect. 3.5), gives similar predictions of the air-gap field in the machine as the models based on conformal mapping and 2D FEM. The analytical expression for the unbalanced force which proceeds from the simplified flux-density model significantly overestimates the force (by about 50–60 % with respect to 2D FEM; practical verification with the test setup was not possible). However, this expression is still satisfactory for motor design purposes—such a model is quite simple and, at the same time, accurate force predictions are rarely needed.

- Manufacturer's data were used to estimate losses in the stator iron core. These losses could not be distinguished from the measurements of overall no-load losses because of the strong impact of the eddy-current losses in the housing. However, the measurement results, in combination with 3D FE simulations, indicated that the losses in the core are somewhat higher than predicted by the manufacturer's data.

- The prediction of the phase DC resistance agrees with the measured values, thereby verifying the model of the DC-conduction loss in the windings. Relatively large deviations of measured resistances of different phases are noticed as a result of an uneven number of turns due to errors in the manual winding process.

- The skin-effect influence on copper losses in the machine is neglected. Indeed, in conductors whose diameter is almost four times smaller than the copper skin depth at the maximum electrical frequency the skin effect hardly has any impact.

- The eddy-current losses in the air-gap conductors of a slotless PM machine can become extremely high unless the conductors are optimized. Analytical models show that the eddy-current losses rise by the fourth power of the conductor-strand diameter while 2D FEM confirms the analytical model predictions. The predictions are, however, neither verified nor refuted by the measurements since these losses could not be separated from accompanying, certainly far more dominant loss factors during the speed-decay tests.

- The proximity-effect loss in the conductors was also neglected as the amplitude of the field of the neighboring conductors in a single conductor strand is expected to be much smaller than the field of the permanent magnet. The measurements did not register any significant influence of the proximity effect.
- No suitable model for representing air-friction loss in the designed machine was found. Rather intricate study would be needed to develop a model for air flow and friction loss in such an unconventional rotor supported by air bearings. According to the speed-decay tests in Sect. 9.4 the models generally used for more conventional machines with slender rotors fail to account for the air-friction loss in the test machine.
- Rotor losses in the motor are neglected due to its large effective air gap and highly-resistive permanent magnet. This assumption is proven to be correct using both 2D FE simulations and locked-rotor tests.

Structural Models

- Modeling of stress in a rotating PM rotor is based on equations which can be found in textbooks on structural mechanics. The suitability of 2D analytical models to represent a PM rotor without magnet-pole spacers as a compound of concentric cylindrical regions is demonstrated by Binder in [75]. This thesis shows that a model which assumes isotropic behavior of the carbon-fiber retaining sleeve by assigning the properties in the direction of fibers to all directions makes almost equally good predictions as a fully orthotropic model of the rotor. Results of these two analytical models were compared to results of 2D FEM and agreement of the models is quite satisfactory.
- Results of 3D FE simulations of the actual rotor showed a greater discrepancy between the predictions of different models, particularly in stresses at corners of different rotor parts; naturally, 2D models cannot account for this. Still, the 3D simulations affirm that the 2D isotropic analytical model is a good basis for the stress calculation and, more importantly, for structural optimization even of such a complex rotor. It is reasonable to expect that such an optimization approach would be even more appropriate for slender rotors with an indeed concentric cylindrical structure. Finally, the very high tangential speed ($\approx$270 m/s) attained with the test rotor confirms validity of the optimization approach.

Rotordynamic Models

- Critical speeds of a high-speed rotor (represented as a cylindrical Timoshenko beam) are correlated with the rotor slenderness and bearing stiffness in Sect. 5.4. This analytical modeling has not been validated in the thesis; however, the results comply with FEM calculations available in literature [137]. The modeling was carried out primarily to show the dependence of flexural critical speeds on machine parameters. These critical speeds are far above the operating speed range of the test motor and validation of these models was not possible. The prediction of the rigid critical speed is confirmed by the tests on the setup (Sect. 9.6).

10.2 Speed Limits of Permanent Magnet Machines

The identification and parameterization of the speed limits of PM machines was set as one of the thesis' main objectives. Here, a short recount of important limiting factors is given.

- The thermal limit is common to all machines. The thermal behavior of a machine depends on power losses that are further dependent on current and magnetic loading, as well as rotational speed. These parameters were theoretically correlated with machine size and rated power in Sect. 2.6. Based on this study it is concluded that, if the cooling method is maintained, it is not possible to gain power density by merely scaling down the machine and increasing its rated speed. Slotless PM machines show the highest capability of speed increase through downsizing; this trend is also confirmed by the empirical study reported in Sect. 2.4.
- In order to prevent high tension in the magnet and ensure the transfer of torque from the magnet to the shaft, high-speed PM rotors are usually enclosed with strong non-magnetic retaining sleeves. In this type of high-speed rotors, the critical, thus limiting stresses are radial (contact) stress at the magnet-iron boundary and tangential stress (tension) at the sleeve inner surface. It was shown in Sect. 4.4 that for an expected operating temperature there is an optimal value of the interference fit between the sleeve and magnet for which both tension and contact limits are reached at an equal rotational speed. This speed can be adjusted by the enclosure thickness so that the theoretical maximum rotational speed is a considerable margin higher than the operating speed.
- Rotation can become unstable in the supercritical regime of a certain vibration mode if rotating damping of the rotor-bearing system affects that mode. Rotors are usually stable in a supercritical speed range which corresponds to rigid-body vibrational modes (with the exception of rotors in fluid journal bearings) and today's high-speed rotors operate regularly in that speed range. On the other hand, rotors which possess some internal damping can easily become unstable in a super-critical range corresponding to flexural modes. Rotors of electrical machines are receptive to eddy-currents, always comprised of fitted elements and often contain materials, such as composites, with significant material damping. Therefore, these rotors are prone to be unstable in flexural supercritical regimes. The first flexural critical speed practically represents the rotordynamical speed limit of an electrical machine.

10.3 Design Evaluation

High-Speed PM Motor

- Loss minimization was taken as the ultimate criterion for the motor electromagnetic design. More precisely, the main intention of the design was to mitigate the overheating of the motor that would result from frequency-dependent losses. That goal is fulfilled when considering electromagnetic losses within the motor: chosen materials, conceptual design and electromagnetic optimization all resulted in a relatively simple machine with very low electromagnetic losses. The slotless design and use of a plastic-bonded magnet and carbon fibers in the rotor resulted in negligible eddy-current losses in the rotor.
- Air-friction loss is rather high given the rotor volume, as expected given the very large diameter of the rotor disc. However, very strong turbulences, influenced partly by the air bearings, very effectively removed the ensuing heat from the rotor and the temperature in the air gap remained quite moderate even after spinning at very high tangential speeds (up to 270 m/s).
- From the efficiency perspective, losses in the aluminum housing are the weakest point of the design; along with air friction, eddy currents in the housing represent the most dominant loss factor in the setup, greatly influencing drive performance and temperature. Although practically inevitable, these losses could have been mitigated with more adequate modeling and dedicated optimization of the stator and housing geometry.
- Unevenly wound coils in motor phases caused an unnecessary increase of the armature field and losses in the stator core. This phenomenon implies that an automatic and highly repetitive winding process is strongly recommended for toroidally-wound machines.
- The rotor structural design facilitated the extremely high tangential speed of the rotor. On the other hand, the design consisting of a few press-fitted elements caused the problem of the recurring unbalance which, eventually, resulted in the bearing crash.
- The plastic-bonded magnet proved to be extremely resistive to eddy-current losses, which is a great advantage for high-speed applications. The relatively small remanent field of this magnet type is suitable for most high-speed machines since their optimum air-gap flux density is usually low. However, the possibility of plastic deformation and creep of the polymer material after long hours of operation has remained a great concern.

Air-Bearings Setup

- Apparently, static air bearings represent a good solution for stable high-speed operation: thresholds of their rotational instability are higher than those of aerodynamic and lubricated bearings. At the same time, very tight tolerances of air bearings impose high precision standards on rotor manufacturing.

- The experience with the motor tests indicates that an assembled rotor prone to unbalance and poorly-damped bearings with very small clearance are not an auspicious match. Such a rotor would require either large damping in the bearings to suppress the oscillations at the critical speed or bearings with large air gaps, such as active magnetic bearings, so that the rotor could benefit from self-aligning. On the other hand, a robust rotor with much tighter tolerances would be appropriate for rotation in air bearings.

10.4 Thesis Contributions

Whole Thesis

- In this thesis, phenomena, both mechanical and electromagnetic, that take precedence in high-speed permanent magnet machines are identified and systematized. The attribute *high-speed* is, for the thesis purposes, defined in Sect. 2.3 and refers to *variable-speed PM machines of small and medium power* (typically below 500 kW) *that have high speed with respect to their power.* A majority of the analyses found in the thesis is applicable to a broader range of (non-PM) high-speed electrical machines.
- The thesis identifies inherent (physical) speed limits of permanent magnet machines and correlates those limits with the basic parameters of the machines. The analytical expression of the limiting quantities does not only impose solid constraints on the machine design, but often also paves the way for design optimization, leading to the maximum mechanical and/or electromagnetic utilization of the machine. Simply put, the analytical expressions indicate where the optimum lies. The most evident example of this can be seen in the rotor structural optimization in Chap. 4.
- The electromagnetic, structural (elastic) and rotordynamical modeling of a (slotless) high-speed permanent magnet machine is presented in the thesis. By juxtaposing the models of different, yet equally important physical aspects of high-speed PM machines, emphasizing relevant machine parameters and presenting those parameters in a logical way, the thesis represents a comprehensive resource for the design of high-speed permanent magnet machines.
- A low-stiffness high-speed spindle drive—a toroidally-wound slotless PM motor with a short rotor—was designed and realized in a practical setup. The design and its practical evaluation offer insight into the merits and drawbacks of using short (gyroscopic) rotor, plastic-bonded magnets and toroidal windings for high-speed machines. Important conclusions are drawn from this design experience and they are listed in Sect. 10.3.

Chapter 2

- Section 2.6 theoretically correlates rated speed, power and size of PM machines, taking into account their physical limits. This theoretical study is supported by an empirical survey of the correlation between rated power and speeds of existing high-speed machines (Sect. 2.4).

Chapter 3

- In Sect. 3.5 an approximate analytical expression for distorted magnetic field in an eccentric-rotor machine is used to determine the unbalanced magnetic force and stiffness of slotless PM machines. The effectiveness of such a model in representing the field in the air gap and unbalanced force is compared to the results of a model based on conformal mapping [108] (the method frequently reported to provide accurate results) and 2D FEM. It is shown in the section that the simplified model gives predictions of magnetic flux density in the air gap similar to the predictions of the conformal-mapping method and 2D FEM and that the prediction of the force is, despite noticeable overestimation, useful and effective for machine-design purposes.
- Using analytical models derived by Ferreira [89, 112], Sect. 3.6.2 distinguishes the dominant causes of copper losses in slotless PM machines. Additionally, the section adapts Ferreira's equation to calculate eddy-current losses in the air-gap conductors of a slotless machine and correlates that expression to a simplified formula reported in literature (e.g. [90]).

Chapter 4

- Structural limits for speed of PM rotors are identified and the limiting parameters (stresses at the rotor material boundaries) are represented in a simple analytical form that clearly indicates optimal geometry of the rotor retaining sleeve. In this way, a relatively simple approach of optimizing the retaining sleeve is achieved; the approach takes into account the influence of rotational speed, mechanical fittings and operating temperature on stress in a high-speed rotor (Sect. 4.4).

Chapter 5

- The dynamic of a rotor-bearings system and its important aspects—stability of rotation and critical speeds—presented in this chapter are fairly known and well-researched in the field of rotordynamics; most models can be found in text-books. The importance of this chapter is that it highlights the effects of those phenomena relevant to rotational stability of electrical machines. In Sect. 5.3 rotor-dynamical stability limits of speed of electrical machines are identified. Section 5.4 analytically correlates critical speeds of a rotor-bearings system using the Timo-shenko beam model; the section shows that the value of the first flexural critical speed (= the limiting speed of an electrical machine) depends solely on rotor slen-derness. The correlation is supported by FE models reported in literature [137].

Chapter 7

- Section 7.2 reports a new spindle-drive concept: a PM motor with a short rotor sup-ported by frictionless (AMB/aerostatic) bearings (together with Kimman [174]).
- In Sect. 7.4.3 exact analytical formulas for the optimization of air-gap conductors in slotless PM machines are derived.
- An innovative design of a retaining sleeve for a short PM rotor is presented in Sect. 7.6; the design consists of a combination of glass- and carbon-fiber retaining rings.

Chapter 8

- The chapter presents a new, open-loop control method for very-high-speed PM motors: an I/f motor controller with reference-frequency modulation. The method is successfully implemented in the test setup.

Chapter 9

- The measurements and FEM presented in the chapter observe certain weaknesses of toroidally-wound machines that have not been emphasized in literature, such as susceptibility to losses in the housing and the necessity of having a flawless winding process in order to avoid excessive core losses.
- The developed motor reached the highest tangential speed of an electrical machine that has yet been reported in academic literature (Sect. 9.6).
- The test setup demonstrates the suitability of aerostatic bearings for very-high-speed operation: a rotor in aerostatic bearings can operate well above twice the critical speed. This renders limits of their rotational stability higher than the limits of aerodynamic and lubricated journal bearings (Sect. 9.6).

10.5 Recommendations

- The representation of the machine in the thesis clearly lacked a thermal model; the designing was, therefore, too cautious and the author refrained from investigating possible thermal limits of the motor in the given setup. The main question which remains unanswered is how to treat the complex fluid/thermal flow in the air gap in a pragmatic way which would be suitable for machine design that would lead to maximum utilization of a high-speed machine. A possible solution that could be explored is to simply bypass the air gap in the thermal modeling—apparently, turbulent air flow at very high speeds is a very efficient means of heat transfer and the effective thermal resistance of the air gap is expected to be very small.
- The modeling of the armature field of a toroidally-wound machine should be improved with respect to the model developed in this thesis. Phenomena that need to be addressed are eddy-current losses in electrically-conductive housing and external leakage of the armature field.
- The presented structural design of the rotor followed the electromagnetic design of the machine. However, the analytical approach for the sleeve optimization lends itself to inclusion into a simultaneous, structural and electromagnetic optimization process. Namely, for defined dimensions of the permanent magnet there is an optimal value of the interference fit and a necessary sleeve thickness for reaching requested rotational speed. If this correlation is considered during the machine design, electromagnetic performance requirements (torque, power, losses) can be achieved with an optimal structural design of the rotor for the desired operating speed.
- The thesis has shown suitability of using plastic-bonded magnets for high-speed machines from the electromagnetic perspective. More research is, however, needed to investigate their suitability with respect to their mechanical properties. In particular, it is important to research whether plastic-bonded magnets can maintain their structural integrity after long hours of operation.
- Practical realization of the proposed short-rotor spindle concept with 5DOF active magnetic bearings would be very interesting. In such bearings, the rotor could benefit from self-aligning at high speeds and, provided that active position control of a gyroscopic rotor is achievable, the concept would result in an extremely compact and powerful spindle drive.

Appendix A

Structural Relationships in a Rotating Cylinder

Governing differential equation for radial displacement in a rotating cylinder (4.7) is derived from Eqs. (4.2), (4.5) and (4.6) in Chap. 4. The first equation is a constitutive structural relation for linear materials—Hooke's law—expressed for the plane stress/strain conditions. The second and third equation represent kinematic and force-equilibrium equation and they will be shortly explained here.

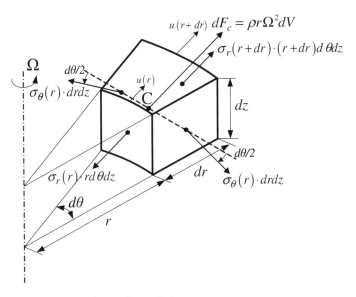

Fig. A.1 A small element of a rotating cylinder

A. Borisavljević, *Limits, Modeling and Design of High-Speed Permanent Magnet Machines*,
Springer Theses, DOI: 10.1007/978-3-642-33457-3,
© Springer-Verlag Berlin Heidelberg 2013

Figure A.1 presents a small element of a axisymmetrical rotating cylinder at a radius r, along with directions and intensities of forces acting on that element. For symmetry reasons, all the structural variables—stress, strain and displacement—are functions of the radius r only, thus independent of the angle θ. Additionally, all shear stress components are zero due to the same symmetry.

As a result of the centrifugal force, all the points in the cylinder undergo a displacement, mainly in the radial direction $(u_r = u(r))$ and certainly not in the tangential direction $(u_\theta = 0)$. Strain represents the relative displacement of the cylinder particles; since no shear components are present, the strain components can be simply calculated as:

$$\varepsilon_r = \lim_{dr \to 0} \frac{u(r+dr) - u(r)}{dr} = \frac{\partial u(r)}{\partial r} = \frac{du(r)}{dr}, \tag{A.1}$$

$$\varepsilon_\theta = \lim_{d\theta \to 0} \frac{(r + u(r))d\theta - rd\theta}{rd\theta} = \frac{u(r)}{r}. \tag{A.2}$$

By referring to Fig. A.1, one can formulate the equilibrium of forces acting on the given element in radial direction:

$$\sigma_r(r+dr) \cdot (r+dr)d\theta dz - \sigma_r(r) \cdot rd\theta dz + 2 \cdot \sigma_\theta(r) \cdot drdz \cdot \frac{d\theta}{2} + \rho r^2 \Omega^2 \cdot drd\theta dz = 0. \tag{A.3}$$

After higher order differential terms in (A.3), the force-equilibrium equation is obtained:

$$r\frac{d\sigma_r}{dr} + \sigma_r - \sigma_\theta + \rho r^2 \Omega^2 = 0. \tag{A.4}$$

Appendix B

One Explanation of Rotor Instability

Figure B.1 displays a cross-section of a Jeffcott rotor which rotates at rotational frequency Ω and, at the same time, whirls around the center of the bearings at frequency equal to the critical frequency $\Omega_w = \Omega_{cr}$. The rotor is subjected to rotating and non-rotating viscous damping. The force from non-rotating damping is proportional to the translational speed of the rotor in the stationary reference frame xy:

$$\overrightarrow{F_{nrd}} = -c_n \overrightarrow{v_{xy}}, \tag{B.1}$$

while the force from rotating damping is proportional to the rotor speed in the vu frame whose origin coincides with the bearing center and which rotates at the rotor frequency Ω, thus:

$$\overrightarrow{F_{rd}} = -c_r \overrightarrow{v_{uv}}. \tag{B.2}$$

(Mind that in a real case rotor displacement r_C is far smaller than the rotor radius.)
Correlation between translational speeds in two reference frames yields:

$$\overrightarrow{v_{uv}} = \overrightarrow{v_{xy}} - \overrightarrow{\Omega} \times \overrightarrow{r_C}. \tag{B.3}$$

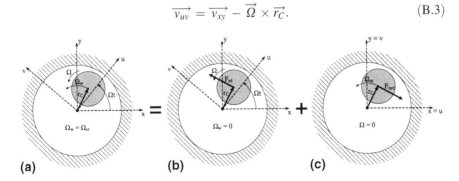

Fig. B.1 Viscous forces acting on a whirling rotor

A. Borisavljević, *Limits, Modeling and Design of High-Speed Permanent Magnet Machines*, Springer Theses, DOI: 10.1007/978-3-642-33457-3,
© Springer-Verlag Berlin Heidelberg 2013

Forces in the example of rotating and whirling rotor (Fig. B.1a) can be obtained as a superposition of forces in two examples: (b) the rotating rotor with a fixed displacement r_C and (c) the non-rotating rotor whirling at the critical rotational frequency.

In the first example non-rotating example has no effect since the rotor is not moving with respect to the stationary frame. Force resulting from the rotating damping is then (according to reference direction in Fig. B.1b):

$$F_{rd} = c_r \Omega r_C. \tag{B.4}$$

In the example of only-whirling rotor (Fig. B.1c) the reference frames coincide and both dampings are active. Force on the rotor in this case is:

$$F_{nrd} = (c_r + c_n) \Omega_{cr} r_C. \tag{B.5}$$

Apparently, the force in the first example is in the direction of whirling while the force in the second example resists the whirling thus is restoring. If the restoring force dominates, the whirl loses the energy and the motion is stable; conversely, if the total force supports the vibration, the whirl will gain energy and increase the amplitude. Therefore, the condition of the rotation stability is:

$$F_{rd} < F_{nrd}, \tag{B.6}$$

which is equivalent to:

$$\Omega < \Omega_{cr}\left(1 + \frac{c_n}{c_r}\right). \tag{B.7}$$

The expression (B.7) is identical to the stability limit analytically obtained in Sect. 5.3.

This intuitive example also demonstrates stability of the backward whirl. Namely, if the whirling has the direction opposite to the direction assumed in Fig. B.1 the non-rotating viscous force will also oppose the whirling and the movement is always stable.

Appendix C

Stator Core Properties

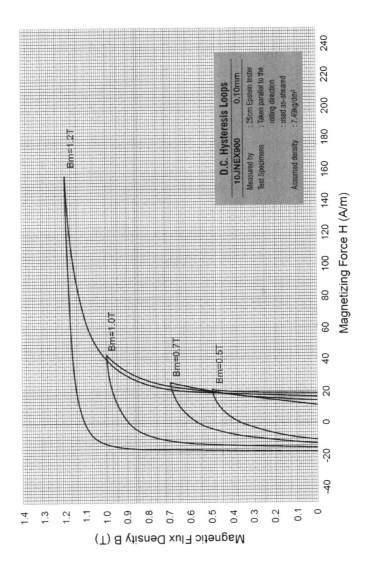

Fig. C.1 Hysteresis loops of 10JNEX900 *Si*-steel [203]

A. Borisavljević, *Limits, Modeling and Design of High-Speed Permanent Magnet Machines*,
Springer Theses, DOI: 10.1007/978-3-642-33457-3,
© Springer-Verlag Berlin Heidelberg 2013

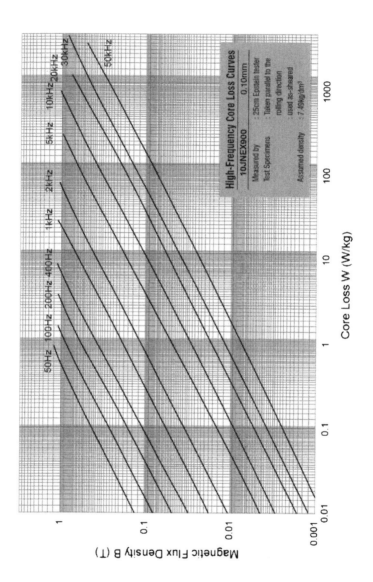

Fig. C.2 Hysteresis loops and loss curves of 10JNEX900 *Si*-steel [203]

Biography

Aleksandar Borisavljević was born in Kragujevac, Serbia, in 1978. He graduated from the Faculty of Electrical Engineering, University of Belgrade, in 2004. After graduation he worked as a researcher at the "Mihajlo Pupin" Institute in Belgrade. In September 2006 he joined the Electrical Power Processing group at Delft University of Technology where he worked toward a PhD degree in the field of high-speed electric drives.

Since November 2011 he has been with Eindhoven University of Technology as well as with Micro Turbine Technologies BV, Eindhoven. At the university he works as an Assistant Professor in the group of Electromechanics and Power Electronics. Both his education and research activities are mainly connected with the design, optimization, testing and control of electric drives.

At Micro Turbine Technology, Aleksandar works as a development engineer focusing primarily on the electromechanical aspects of Combined Heat and Power systems.

A. Borisavljević, *Limits, Modeling and Design of High-Speed Permanent Magnet Machines,* 215
Springer Theses, DOI: 10.1007/978-3-642-33457-3,
© Springer-Verlag Berlin Heidelberg 2013

Printed by Publishers' Graphics LLC